SANITIZING MOSCOW

HISTORY OF THE URBAN ENVIRONMENT
Martin V. Melosi and Joel A. Tarr, Editors

SANITIZING MOSCOW

Waste, Animals, and Urban Health in Late Imperial Russia

Anna Mazanik

UNIVERSITY OF PITTSBURGH PRESS

Published by the University of Pittsburgh Press, Pittsburgh, Pa., 15260
Copyright © 2025, University of Pittsburgh Press
All rights reserved
Manufactured in the United States of America
Printed on acid-free paper
10 9 8 7 6 5 4 3 2 1

Cataloging-in-Publication data is available from the Library of Congress

ISBN 13: 978-0-8229-6772-9
ISBN 10: 0-8229-6772-3

Publisher: University of Pittsburgh Press, 7500 Thomas Blvd., 4th floor, Pittsburgh, PA 15260, United States, www.upittpress.org
EU Authorized Representative: Easy Access System Europe, Mustamäe tee 50, 10621 Tallinn, Estonia, gpsr.requests@easproject.com

To my grandparents

CONTENTS

Acknowledgments

xi

Notes on Transliteration and Dates

xiii

INTRODUCTION. Russia's Sanitary City

3

Part I. A Quest for Clean Modernity

CHAPTER 1. Discovering Moscow's Dirt

13

CHAPTER 2. Making a Sanitary City

31

Part II. Water, Waste, and Technologies of Sanitation

CHAPTER 3. The Sewage System:
European Symbol, American Design

53

CHAPTER 4. What Happened to Waste?

70

CHAPTER 5. The Conundrum of Industrial Waste

85

Part III. Animal Bodies for Human Health?
Livestock and the Sanitary Project

CHAPTER 6. Feeding Moscow:
Livestock Trade and Medicalization

101

CHAPTER 7. The Killing Factory

115

CHAPTER 8. Civilized Slaughter:
Animal and Human Welfare

132

Part IV. A Paradox of the Sanitary Project: Children in Moscow

CHAPTER 9. **A Deadly City for Children**

149

CHAPTER 10. **Healthy Schools in a Deadly City**

173

CONCLUSION. **Across the Divide**

201

Notes

211

Bibliography

235

Index

257

ACKNOWLEDGMENTS

I am often surprised how fortunate I am to have so many people who helped and supported me during the years of working on this book. First, I would like to thank Alexander Shevyrev from Moscow State University who many years ago introduced me to Moscow urban history. This is where my long research journey started.

This book grew out of my doctoral dissertation written at Central European University in Budapest and owes a lot to the unique intellectual community that existed there in the late 2000s and 2010s. I would like to specifically thank Alexei Miller, who supervised that dissertation, Maria Falina, Anastasia Felcher, Gabor Gyani, Karl Hall, Yulia Karpova, Uku Lember, Goran Miljan, Ohad Parnes, Markian Prokopovych, Alfred Rieber, Mate Rigo, and Irina Savinetskaia.

I owe deep gratitude to Daniel Todes and Graham Mooney from the Institute of the History of Medicine at Johns Hopkins University. Without their guidance in the history of public health this book would have been completely different.

Many scholars in several countries discussed my project with me and offered valuable suggestions. Special credit goes to Andrei Davydov from the Central State Archive of the City of Moscow, whose publications and comments guided me through the archival holdings. I would like to thank

Acknowledgments

Alexander Friedrich, Mikael Hard, and Dieter Schott from TU Darmstadt, where I wrote the first draft of the part on the slaughterhouse. The project on interurban knowledge exchange in Southern and Eastern Europe, organized by Eszter Gantner, Heidi Hein-Kirchner, and Oliver Hochadel, greatly influenced the way I thought about my book. Anastasia Fedotova, Emily Gioielli, and David Moon kindly read the introduction and the book proposal.

My time at the Rachel Carson Center in Munich was essential for completing this book. I would not have been able to finish this book without the support of Anna Pilz and the writing group she led. Rob Gioielli and Matthew Klingle discussed the book's structure with me, and their suggestions had a considerable impact on how the text is organized. Tom Lekan read the part on sewerage and provided many useful comments. The conversations with many other fellows and colleagues at the RCC are also reflected in this book.

My research has been supported by scholarships and grants from Central European University (Budapest), DAAD (German Academic Exchange Service), the Technical University of Darmstadt, and the Herder-Institut for Historical Research on East Central Europe (Marburg) as well as the Pontica Magna Fellowship at the New Europe College–Institute for Advanced Study (Bucharest).

Finally, I would like to thank my family for their enormous support over the many years of my work on this project. My family in Moscow hosted me during my long research trips to Russia (which in recent years also meant babysitting my children while I was in the archive or library). They scanned and photographed documents for me and took care of me in many different ways. My German parents-in-law helped with childcare and also read the drafts of some chapters. But my greatest gratitude goes to my husband, Brendan Röder, a wonderful partner and co-parent, who has been in the trenches of academia with me all this time and read every line of this book at least twice.

NOTES ON TRANSLITERATION AND DATES

I use the Library of Congress transliteration system except for some personal names and toponyms. German and English names appear in their original spelling in the text (Erismann, Williams, Witte) but are transliterated in the notes and bibliography when the Russian-language publication is referenced (Erisman, Vil'iams, Vitte).

Dates follow the Julian calendar ("old style"), which was used in Russia until 1918. It was twelve days behind the Gregorian calendar in the nineteenth century and thirteen days behind in the twentieth century.

SANITIZING MOSCOW

INTRODUCTION

Russia's Sanitary City

In summer 1917, Charles-Edward Amory Winslow, the first head of the Yale Department of Public Health and one of the leading figures in the American public health movement, visited Russia as part of an American Red Cross medical mission. The country was in the middle of a revolutionary crisis, which would bring the Bolsheviks to power several months later; it was also three years into a devastating war, struggling with the food shortages, inflation, violence, crime, chaos, and displacement that were entailed.[1] Upon his return to the United States, Winslow wrote a report on health administration in Russia, a surprisingly positive account of both the past achievement and the current work of Russian public health:

> In the first place, one is impressed with the possibilities of the numerous advisory boards, made up largely of active employees, with which both zemstvo [rural governments] and municipal executives are surrounded. Such organizations must often prove cumbrous and time consuming, but they tend to favor initiative and esprit de corps on the part of the staff. It is interesting to note that even before the revolution Russia was in this respect in position to give a lesson in democracy to the rest of the world.
>
> The great strategic point in the Russian health situation is, however, the remarkable development of social medicine along curative lines and the

consequent close connection between curative and preventive work. Russia . . . has already developed the State care of the sick to a point of which we are only beginning to dream.²

Although clearly aware of and noting Russia's health problems, such as high rates of some communicable diseases and extreme infant mortality, Winslow considered the health administration well organized, adequate for the country's needs, and effectively functioning despite the war and revolution. He found Russian engineering "solid and successful," the personnel knowledgeable, and medical statistical bureaus "better equipped with funds and highly trained professionals than our own."³

The Red Cross mission traveled across the country, starting in Vladivostok and moving toward European Russia, and Winslow visited Petrograd (Saint Petersburg), the imperial capital, and Moscow, which would become the new capital the following year. Moscow, according to Winslow, was the leader in Russian municipal health work, just as the government of the province of Moscow was the leader in rural health. Winslow left a detailed description of the municipal sanitary organization, noting many similarities with the United States and commending the city's effective measures against water-borne and insect-borne diseases, modern clinics, adequate school sanitary inspection, well-equipped laboratories for food and water control, an "elaborate disinfection station," and "admirable municipal lodging houses." Two aspects of Moscow's public health and sanitary system appeared to him particularly innovative. Although Winslow again noted the appalling death rates among young children, he described the infant welfare station at one of the municipal hospitals as the "most perfectly equipped plant for this purpose" that he had ever seen. But it was the Moscow sewage system that left the biggest impression on him and especially the experiments with sewage treatment carried out there in 1917, which were, according to Winslow, "probably the most extensive and important sewage treatment studies being conducted in the world" at that moment.⁴

Winslow's account is very different from the common tropes found in modern travel literature on Moscow and its sanitary condition, where the city is described as filthy, exotic, poor, technologically, institutionally, and scientifically backward, and essentially very different from a "Western" city. Instead, Winslow offers an image of modernity, rationality, effective management, and innovation; he emphasizes not the difference but the parity and commonality with "Western" or at least American urbanism. Neither does his report fit with the texts by Soviet authors who claimed that all improvement in urban public health and sanitation started only with the new regime.⁵

It is also at odds with existing narratives produced by historians. Late imperial Moscow has been the subject of several historical studies, though no work has dealt specifically with its public health, environment, and sanitation. Such topics, however, do inevitably come up in more general studies of the city's history, which produce a picture of "urban crisis," failure, and backwardness, and of Moscow as being the "unhealthiest" or "deadliest" city in Europe.[6] When Donald Filtzer, the author of perhaps the most thorough anglophone study of public health and urban sanitation in Soviet Russia, needed to historicize the experience of postwar Soviet cities, he referred to sanitary reforms in Western European and American cities of the nineteenth and early twentieth century and not to the cities of the Russian Empire, as if no similar processes or discussions of urban infrastructure, public health, and sanitation ever took place there.[7]

Yet, in late imperial Russia, the questions of health and sanitation in the context of rapid social change were a subject of intense scientific and broader public debate and a highly contested arena of policy. In the last decades of the nineteenth century, Russia's governmental institutions, particularly at a local level, promoted public health reform as an essential component of the modernizing agenda. The centrality of health and sanitation in the vision of modernity led Russia's educated elites to see their mission as making the country and its population healthier in order to bridge the perceived gap with "advanced" Western societies. Historians of the late imperial period have shown that these questions were crucial for the professionalization of the Russian medical community; they were important drivers of local civic activism and political mobilization and, by the early twentieth century, also a sphere of interest for the central imperial government that would then culminate in the creation of the Main Administration of Public Health (Glavnoe Upravlenie Gosudarstvennogo Zdravookhraneniia), Russia's equivalent of a public health ministry, in 1916.[8] And, as in Western cities, these evolving public health concerns served as a justification for municipal intervention and for the introduction of social reforms, public services, and new infrastructures and influenced urban experience and urban environment in a multitude of ways.

Nineteenth- and early twentieth-century urban public health, environment, and infrastructural development in Western Europe and North America—and the emergence of what Martin Melosi has called the "sanitary city"—are the subject of a rich scholarship that has, in many ways, shaped the research behind this book. Urban, environmental, and public health historians have explored how and to which extent urban health landscapes and responses to sanitary problems were influenced by experts, politicians, and the public, by transnational scientific ideas and local politics, by the changing understandings of disease and the appearance of new technologies, and by all of these blending with more established cultural practices. Environmental historians discuss the impact of urban sanitary services, technologies, and public health measures on the more long-term patterns of urban development, on the environments in and outside the cities, and (increasingly so) also on the nonhuman entities of urban ecosystems.[9]

Although historians of late imperial Russia have studied public health on the national level, in relation to rural policy or in the context of specific diseases, urban public health has been little investigated in the anglophone scholarship.[10] The problems of urban pollution and sanitation have also been little studied by environmental historians of Russia, despite the fact that it has been a firmly established topic of environmental historiography on other regions.[11] Russian environmental history is clearly on the rise, but the imperial period has received comparatively modest attention beyond the prelude to the Soviet-focused studies and then mostly in the context of water, steppe, or forest environments of imperial borderlands rather than in cities.[12]

In the Russian language, there exists considerable scholarship on late imperial urban health and sanitation, including some on Moscow.[13] However, these studies are usually written from a perspective of local or institutional history, and their authors do not engage with the broader social and environmental impacts of sanitary policies or the transnational context of medicine, infrastructure development, and urban reform.[14] At the same time, there is growing interest in the history of imperial biological and environmental sciences, but there has been little research on how those sciences were integrated into the practice of urban health.[15] The nexus of urban, environmental, and public health histories of late imperial Russia remains underexplored.

Among Russian cities, Moscow, the second metropolis of the empire and Europe's sixth largest city in 1900, introduced the most extensive program of sanitary reforms. Those reforms created a new sanitary regime—the Russian version of the "sanitary city"—based on the medicalization of urban politics and the municipalization of public health when urban problems were increasingly interpreted through the lens of medical science and managed by medical professionals and where rapidly expanding sanitary services and health provisions were designed and delivered by the municipality rather than by the central state or private businesses. Despite its many shortcomings, controversies, and inconsistencies, this was an ambitious and innovative program with profound social and environmental effects, and this dynamic transformation is obscured by dry references to grim mortality and morbidity statistics that often guide the evaluation of such reforms in more general works on Moscow's urban history.

It is these reforms that are the subject of the present study. How and why did Moscow elites understand and engage with sanitary problems? Why did it happen at that particular moment? How did their responses to those problems affect the city, the lives of its inhabitants, and the environments around them? Who benefited from sanitizing Moscow, and who was excluded? How can one reconcile the sanitary reforms with Moscow's stunning mortality rates? Why was Russia's sanitary city so deadly?

One goal of the book is to revise the image of the sanitary reforms as being logical answers to the objective challenges of urbanization, industrialization, and the deteriorating environment and health. Instead, I show that, in Moscow, the reasons for the sudden public concern with pollution and sanitation were not so much biological as social and political. The interventions in public hygiene were determined not by the incidence of a particularly dangerous disease or the scale of an ecological problem but by their power to attract attention and to fuel relevant political debates over modernization, social progress and Europeanness. That is not to downplay the objective challenges that the Moscow government was facing. The critical, in fact, life-and-death urban problems in late imperial Moscow were very real, and so was the impressive program of sanitary reforms implemented by the municipality. However, as I reveal in this book, that program failed to address some of the most crucial and deadliest aspects of urban life, and it is impossible to understand this mismatch without considering the cultural factors and politics behind it.

Another goal is to place Moscow sanitary reforms into a broader transnational context of modern public health and urban development. In Russian studies, late imperial municipal public health is often viewed as a derivative and inferior version of the innovative system of community medicine pioneered by the rural zemstvos with their free health care and the emphasis on preventive work. Yet many of the challenges facing Moscow in the late nineteenth century were unknown to the zemstvo but were similar to those tackled in other rapidly growing cities across the globe. Although strong intellectual and direct personal ties between Russian rural and urban community medicine are undeniable, Moscow was also firmly involved in the networks of interurban exchange that historians have referred to as "transnational municipalism."[16] I argue that the competition and exchange with other cities was essential for shaping the politics of public health and sanitation that were envisioned as an instrument in overcoming Moscow's perceived "backwardness." These overlooked and unexplored transnational connections are crucial for understanding the trajectory of Moscow sanitary policies, which by the turn of the twentieth century significantly diverged from the zemstvo model toward a new approach based on technological solutions to sanitary problems.

Expanding the focus beyond the lens of Russian community medicine allows us to see whether and in which ways the sanitary movement in Russia was similar to such movements in the West. There has been considerable debate on the influence that the changing scientific epistemology, bacteriological discoveries, and the appearance of laboratory medicine had on the practice of public health in the late nineteenth century. In particular, historians working on Britain and France argue against the validity of the notion of "bacteriological revolution" as a radical and clear-cut shift that transformed every aspect of medical science and practice. Instead, they advocate a longer, more gradual and subtle evolution and a more "synthetic" vision of the relations between the new bacteriological knowledge and the older patterns of medical and sanitary work.[17] In the histories of Russian public health, however, bacteriology has been given a rather small part. Although scholars have noted the early reception of bacteriology by (some) Russian scientists, the focus on the zemstvo community medicine in the countryside led to downplaying the influence that bacteriology had on the public health work, at least before the first decades of the twentieth century.[18] By shifting attention to the urban context, in this book I offer a different, more nuanced perspective on the relations between bacteriology and sanitation and the role that the rise of the germ theory of disease and laboratory medicine played in the local practice of public health in Russia.

Finally, I want to emphasize that public health and urban sanitary reforms were themselves powerful forces in shaping the environment and human relations with the nonhuman world, that their impact went beyond the human and influenced other living beings. Imperial and urban reformers primarily had human health in mind, but those reforms were not about humans alone. Russian medical professionals placed human health inside an intricate web that connected natural, social, and political conditions, the physical environment, and animal health. By establishing new practices of disease prevention, setting new standards of cleanliness and construction norms, changing patterns of waste disposal, water usage, and dietary recommendations, the public health movement and specific sanitary reforms changed not only Moscow's urban space but also the way Russians thought about and dealt with environments outside the city. Although the human experience remains central to my project, I want to show how public health measures affected human interactions with (other) animals, ranging from extermination, exclusion, and relocation to the provision of advanced veterinarian health care.

The book is divided into four parts. In the first part, I discuss why Moscow elites suddenly became interested in the environment and public health and provide the historical setting for the case studies that follow. In chapter 1, I explore how disease, pollution, and the environment became a matter of public concern in the 1870s and 1880s, and I examine the political, social, and scientific context of Moscow's reforms. In chapter 2, I offer a chronological overview of the Moscow sanitary project and highlight its evolution from the zemstvo model to a model based on specialized medical care and complex technological infrastructure. I describe the scope and key stages of the sanitary reforms and the discourses and politics behind them, including the changing relations between the imperial administration and the municipal government.

In the second part of the book, I study the problems of waste and water pollution with a focus on the Moscow sewage system (1887–1898), the largest and the most expensive infrastructural reform undertaken by the Moscow government. In chapter 3, I discuss competing projects of the Moscow sewage system in the context of transnational interurban exchange and the eventual choice of the separate system modeled on George Waring's design for Memphis, Tennessee. I show how striving for European metropolitan modernity together with the rise of the germ theory paradoxically led Moscow to adopt a design from a much smaller American city that was created by a supporter of the miasmatic theory of disease. In chapter 4, I explore what happened to waste in the Moscow sewage system, whose waste went in, where it ended up, and what it meant for the environments and peasant communities of the Moscow suburbs. In chapter 5, I investigate how the commissioning of the Moscow sewage system transformed the public debate, legal treatment, and policy toward the industrial pollution of waterways in the wider industrial region of Central Russia.

In the third part of the book, I examine the medicalization and industrialization of animal slaughter through the case of the Moscow abattoir (1886–1888), the project in which Moscow reformers took particular pride. In chapter 6, I study the origins of the livestock slaughtered in Moscow. I tell the hitherto unknown history of Russian cowboys who were driving cattle to Moscow markets from the steppe regions north of the Black Sea, the Caspian, and even from Central Asia, and I discuss the part that medical considerations and policies played in shaping this process. In chapter 7, I analyze slaughterhouse reform in Moscow, the motivations behind it, the built environment of the abattoir, as well as the medicalization of animal slaughter and its influence on disease control in Moscow and the distant stock-raising regions. Drawing on the rich correspondence between the abattoir's management and the Russian Society for the Protection of Animals, in chapter 8 I unravel the approaches to animal rights and welfare in tsarist Russia and show that the debates about the well-being of animals and humans were tightly linked.

In the last part of the book, I investigate what I consider to be a major paradox of the Moscow sanitary project—the problem of children's health in the city—through the lens of two contrasting cases. In chapter 9, I examine the high infant mortality rate in Moscow, which played a dominant role in the city's general mortality statistics and, therefore, in the overall evaluation of the Moscow sanitary project. I discuss the social and environmental factors that contributed to this extreme mortality and the measures the municipal government did and did not introduce to control it. Finally, in chapter 10, I look at the emergence of school hygiene and school sanitary inspection in Moscow (1889) at the same time as primary education was dramatically expanding in the city. By analyzing the daily work of school sanitary doctors in Moscow, I explore how medical professionals shaped the school environment and, through it, the experience of schooling and urban childhood more broadly.

PART I

A Quest for Clean Modernity

CHAPTER 1

Discovering Moscow's Dirt

In 1879, a municipal committee in Moscow stated that "Moscow has long ago become the model of pollution and negligence. Despite its favorable topographic conditions, the abundance of water sources, considerable vegetation, the city became unhygienic in all aspects because of the indifference of the residents to the public interest."[1] In 1884, the year when the Moscow sanitary organization was established, one of the city's satirical magazines welcomed the new institution with the following words: "We feel sincerely sorry for the newly created sanitary doctors, who, with a zeal worthy of a better cause, are forced now to play a comedy of inspection. We would like to be mistaken, but, judging from the experiences of the past years, we can hardly believe in the good results of the sanitary campaigns against Moscow's dirt. This dirt is primordial, original, accumulated for centuries; what can the weak hands of sanitary doctors do about it?"[2]

What ties these two quotations together, and many other similar quotations of that time, is the belief that Moscow was dirty, very dirty. At the turn of the 1880s, the Moscow public suddenly discovered that their city was polluted, filthy, unhealthy, and disorderly and (more important) had been that way for a long time. Although "Moscow's dirt" had presumably been there for a while, the widely shared perception of dirt and pollution as a problem was new.

Before that, the Moscow public was not too concerned about their city's dirt. As historian Alexander Martin argues, in the first half of the nineteenth century Moscow with its small houses and abundant gardens cultivated the myth of being a healthy and harmonious place and enjoyed a sense of superiority over the West European cities that suffered from overcrowding and social conflict. Although the Moscow public was aware of dirt and fetid odors, they did not invoke any images of decay or danger, but rather those of closeness to nature, patriarchal mores, and social harmony. In Martin's interpretation, the silence of the Russian public about the everyday filth and stench of Moscow was a way of affirming the ancien régime values and vitality of the country's social order.[3]

Why did the Moscow public suddenly become concerned with the dirt, pollution, and sanitation in their city in the 1870s and 1880s? My main argument is that this concern was connected not only to the objective challenges of demographic and industrial growth, densification of urban space, and pressure on natural resources but also to broader social, political, and scientific transformations. What changed was not only the level of pollution or the risk of getting sick in the city but the people who looked at the city—where, how, and why they looked, how they explained what they saw, how they evaluated it and compared it to other places, and what they could do about it. In this chapter, I discuss the factors that contributed to that change: the social transformation of the city following the Great Reforms, the appearance of a more powerful municipal government, the establishment and institutionalization of scientific hygiene and bacteriology, the development of community medicine, the growth of transnational municipalism, and the beginnings of a new type of city politics in which public health and the reorganization of the urban environment would play a major role.

Urban Growth, Social Transformation, and City Government

The second half of the nineteenth century was an era of modernization and rapid growth for many cities around the world. Russia was no exception, and among its cities Moscow, though not unique, was a striking example of this process (see figures 1.1 and 1.2). Moscow possessed many distinctive features of Russian urbanization, but it was not a typical town. The sheer scale of the population set Moscow aside.

The size and metropolitan claims of imperial Moscow put it in the same league as major European cities. By the turn of the twentieth century, among European metropolises Moscow was ranked sixth in population and was the biggest city without a capital status. The crucial point in the growth

FIGURE 1.1. Moscow. View from the Ivan the Great Bell Tower. From *Moskva. Vidy nekotorykh gorodskikh mestnostei* (Moscow, 1884). National Electronic Library (Russia).

FIGURE 1.2. Moscow. View from the Church of Saint Simeon Stylites. From *Moskva. Vidy nekotorykh gorodskikh mestnostei* (Moscow, 1884). National Electronic Library (Russia).

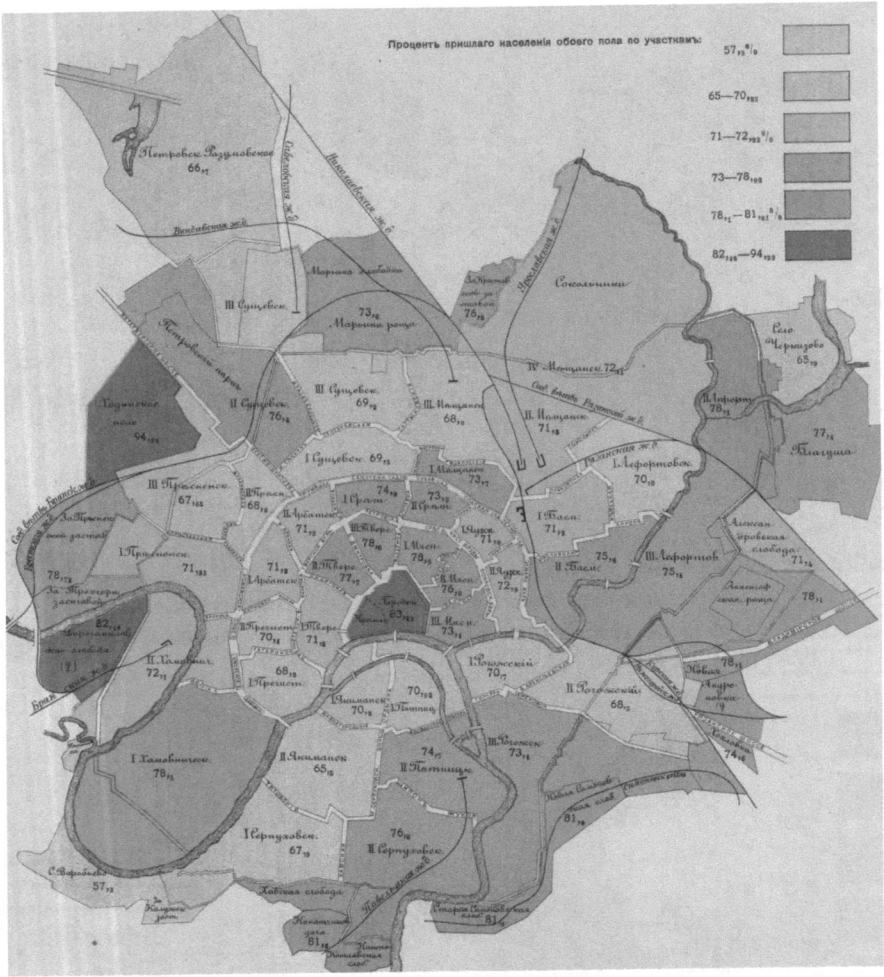

FIGURE 1.3. Plan of Moscow with the percentage of immigrants in each district in 1907. From *Statisticheskii atlas goroda Moskvy* (Moscow, 1911). National Electronic Library (Russia).

of Moscow was connected to the abolition of serfdom in 1861. In the first decade after the liberation, the city's population increased by half, reaching 602,000 in 1871. The pattern of growth continued in the following decades, and Moscow had 1,043,000 residents in 1897, 1,346,000 in 1907, and 1,612,000 in 1912.[4]

Most of the migrants belonged to the peasant estate and came from the central provinces of European Russia, a region that suffered from overpopulation and a lack of land. The peasants were forced to look for alternative means of subsistence and going to the city in search of wages (*otkhod*) became a common solution. The abolition of serfdom intensified the peasant migration to Moscow due to the facilitation of the process and due to the necessity of redemption payment for the land after emancipation.[5]

In the last decades of the nineteenth century, Moscow was a true city of immigrants. The fact that immigration was the dominant factor of urban growth was not something uniquely Russian. Yet in the 1880s, among the major European cities, Moscow had the greatest proportion of immigrants—about three-quarters—and still held this position twenty years later (see figure 1.3). The ratio of immigrants was particularly striking among the labor force—in 1902, a mere 12 percent of the entire active male population were born in Moscow.[6]

What is peculiar about the Russian urbanization experience is the transient character of migration and the strong ties of migrants to their village of origin. The common survival strategy was the combination of agricultural labor and additional wages in the city within the same family. *Otkhod* remained a predominantly male phenomenon: husbands and sons went off to the city while wives together with children and older family members remained working in the village. The gender composition of Moscow was a good illustration of this pattern. In London, Paris, Berlin, and Vienna, women outnumbered men, whereas Moscow had only 700 females per 1,000 males in 1871, the figure slowly increasing to 755 in 1897 and to 767 in 1902.[7]

However, it was not only the tremendous influx of peasant migrants that changed the city. The elites changed too, both in terms of their social composition and their involvement in urban politics. Moscow's industrialization, predominantly in textile production, made it one of the key centers of the emerging Russian bourgeoisie. The old nobility, deprived of their main income after the peasant reform, were challenged by the new elite, the nouveaux riches whose supremacy was built on their success in the capitalist economy. "Merchant Moscow" was the late-nineteenth-century nickname of the city. Pavel Buryshkin, a merchant himself, who popularized this nickname by using it as a title for his famous memoirs, chose the following quotation to describe the role merchants played in Moscow life: "You cannot take a step in Moscow without a merchant. . . . Everything outstanding in Moscow is either in the hands of a merchant or under his feet. He has the best houses and carriages, the best paintings, lovers, and libraries. Whichever institution you look into, you inevitably meet a merchant there, wearing a suit, with an English pleat, speaking French, but still a merchant."[8] Those "Moscow merchants"—many of them originally of the traditional Russian Orthodox or Old Believer background and Slavophile worldview, who within several generations went all the way up from the lower classes to be industrialists and bankers—played a decisive role in the post-reform municipal politics, both through their direct participation in elected government and through generous donations that helped finance various municipal initiatives, including education, welfare, and public health.[9]

Moscow, the home of Russia's biggest university as well as university-like higher courses for women, was also a magnet for the Russian intelligentsia, drawn particularly from the sons and daughters of the two educated social groups of the Russian Empire, the gentry and the clergy, but later on also increasingly from humbler social background. The young generations of the reform and post-reform eras shared a faith in science, social and technological progress, modernization, and westernization. Many believed that scientific advances, the reformed government, and the reorganization of society according to the principles of reason would change Russia and help improve the lives of its people. They joined the ranks of scientists, physicians, engineers, lawyers, statisticians, teachers, and journalists who would form the professional community of late imperial Moscow. Although embedded in Russia's distinctive sociopolitical realities, they were at the same time well connected with corresponding professional circles in Western societies.[10] It was this social stratum that provided expertise to the city government and employees to the expanding municipal services.

The peculiar alliance of these new entrepreneurial and professional elites, heterogeneous as they were, shaped urban reforms in the last decades of the nineteenth century. Moscow was governed by the Moscow City Council (Gorodskaia duma)—a large policymaking body that elected a small executive board (Gorodskaia uprava) and the mayor. According to the Municipal Statute (Gorodovoe polozhenie) of 1870, suffrage was given to Russian citizens who paid taxes to the city, either from real estate property or in the form of commercial or industrial fees. Only men over the age of 25 could personally take part in the elections, but women and men under 25, paying the necessary taxes, could express their will via warrants. The inequality of electoral qualification was expressed by the curial system, which divided the electorate into three groups depending on their tax contributions. Each group collectively paid the same amount of taxes to the city and elected the same number of deputies to the council, thus privileging wealthy entrepreneurs and property owners. All those who rented an apartment in the city and who were not engaged in commercial activity—that is, large groups of professionals, civil servants, and workers—were excluded from the electorate.[11]

There were two important aspects of Moscow municipal politics in the last third of the nineteenth century that distinguished it from Western and

Central European or North American cities of comparable size. First, the independent municipal rule was new in Russia. Despite several controversial attempts to develop urban self-government institutions since the eighteenth century, Russian cities remained both politically and financially weak and subordinated to the central state, which initiated and carried out most large urban improvements. It was not until the municipal reform of 1870 that cities were given the right to "act independently," to form and spend their own budgets, to impose taxes, freely use municipal properties, take out loans, and regulate urban life through introducing rules that city dwellers had to follow.[12] The new municipal statute empowered municipalities to put forward extensive urban reforms, but the novelty of the field meant that this process was inconsistent and highly dependent on individual initiative and the influence of specific experts.

Second, as Russia remained an autocratic state, municipalities in the cities together with zemstvos in the countryside were the highest elective political bodies until the appearance of the State Duma (Gosudarstvennaia duma) following the 1905 Revolution. This meant that municipal autonomy had its limits, and urban reformers had to stay in constant dialogue—and sometimes enter into conflict—with central and local administration. This also meant that municipal politics were not about the city alone. Although the activity of the Moscow reformers was a local enterprise, it had a much wider resonance in the country where municipalities and zemstvos served as a kind of test model of an elected liberal government. As the responsibilities of municipalities and zemstvos were legally limited to the "apolitical" realm of infrastructure, beautification, welfare, sanitation, and public health, these spheres provided a rare legal arena for civic activism, political contest, and reformist ambition.

Hygiene, Germs, and the New Science of Public Health

What was changing in the second half of the nineteenth century was not only Russia's social and political life but also the world of medicine, including its institutional settings, its status in society, and the scientific understanding of disease. During the era of the Great Reforms, Russia saw a concerted effort to promote science and medicine—and scientific medicine—and to equip the country with more and better-trained physicians and better instruments to combat disease. This effort, including the university reform of 1863, led to a substantial increase in state funding for medical institutions, revision of curricula, and the expansion of faculties to include scientists trained in Western Europe. The law of 1876 entitled "Measures to Increase the Number of Physicians in the Empire" provided better financial aid to students and attracted more young people from poorer families to the medical faculties, where enrollments rose rapidly and remained high for the following decades. At the same time, new local governments created more salaried positions for physicians, making public health work a viable career for medical graduates and also a career that promised greater professional autonomy.[13]

One of the many results of these changes was the institutionalization of hygiene and its development as the science of public health. Until the mid-1860s, hygiene was a secondary subject within the field of legal medicine, but many members of the medical establishment believed that it should be granted a better place within the university curriculum. In 1865, the Medical Department of the Ministry of Internal Affairs started publishing the *Archive of Forensic Medicine and Public Hygiene* (*Arkhiv Sudebnoi Meditsiny i Obshchestvennoi Gigieny*), a professional journal that was sent for free to all state-employed physicians across the empire and would soon emerge as an important forum for discussing the social questions of medicine and public health. Simultaneously, the separate chair of general, military, and maritime hygiene was established at the Military Medical-Surgical Academy in Saint Petersburg, acknowledging the role of hygiene for maintaining the effectiveness of the army after the debacle of the Crimean War. In 1868, the Ministry of Public Instruction and the Ministry of Internal Affairs agreed that hygiene deserved a separate chair at the universities as well. It took, however, more than a decade until Moscow University filled that new chair by inviting Friedrich Erismann, a scientist who would play a major role in the development of hygiene as a discipline, and connecting hygiene to Russia's community medicine.[14]

Friedrich Erismann (known in Russia as Fyodor Fyodorovich Erisman) was initially trained as an ophthalmologist and earned his doctorate at the University of Zurich. There he became familiar with the Russian émigré community (including many radical and revolutionary youth) and met his future wife, Nadezhda Suslova. The encounter would change his life and bring him to Russia. The daughter of a serf, a feminist, a member of the revolutionary organization Land and Freedom, Suslova was then the only female medical student officially enrolled at the University of Zurich. She would be the first Russian woman to get a doctorate in medicine. In this period, Erismann became interested in social democratic politics and joined the Swiss section of the First International, although it is unclear if this interest developed directly under Suslova's influence or if their marriage was itself a result of their shared worldview.[15]

In 1869, Erismann followed Suslova to Russia. While working as a private practitioner in Saint Petersburg, he started investigating environmental factors in disease causation, first in the field of his initial expertise, studying the influence of school arrangements on the development of myopia, and then moving on to explore the housing conditions of the poor. To expand his knowledge in hygiene and sanitation, he went to Munich to study with the hygienist Max von Pettenkofer. After the Russo-Turkish War of 1877–1878, when Erismann was involved in the disinfection works at the front, he was called to work for the Moscow zemstvo and, from then on, was actively involved in Russia's expanding community medicine. By the time he was invited to lecture at Moscow University in 1882, Erismann was the author of several programmatic and comprehensive works on hygiene that defined this science, its subject matter, tasks, methods, and its role in society: a three-volume *Handbook of Hygiene* (1872–1878), *Professional Hygiene, or the Hygiene of Mental and Physical Work* (1877), *Popular Hygiene* (1878), an essay "Organization of Public Hygiene in Russia" (1876) as well as several shorter texts on specific aspects of hygiene and sanitation. In 1884 Erismann received his own chair within the medical faculty and created a hygiene laboratory that in 1891 would become the Institute of Hygiene.[16]

Erismann was enormously influential in Russian professional circles. In addition to being a charismatic speaker, a prolific writer, and a successful publicist, he was one of the founders of the country's main professional medical organization, the Pirogov Society of Russian Physicians, and a leading spirit behind its focus on preventive and sanitary work and the broad social framing of its agenda. While he was a professor of hygiene (which now became a mandatory subject of the medical curriculum), Erismann trained a new generation of Russian physicians and used his professional authority to support the career of those who shared his views.[17] This was helped by the fact that Moscow University had Russia's largest medical faculty and produced about one-third of all new medical graduates.[18] It is worth looking in more detail at his interpretation of hygiene and sanitation and their relations to germs and the environment.

What was this discipline of hygiene that Erismann was promoting? In his words, hygiene was a science that studied "all those natural conditions and factors of social life that in one way or another contribute to the disturbance of physiological functions of the human organism and thus influence the mortality and morbidity of the population." The main goal of all research in hygiene, according to Erismann, was to find the laws that govern human health. To explore these laws, hygiene had to expand its focus beyond internal physiological processes and to look at the relations between the human body and its environment, including both natural and man-made factors. When speaking in front of the university students in the late 1880s, he described hygiene as a new, young, flourishing science and said that scientific hygiene as a discipline had been born "fifteen–twenty years ago" (that is, in the late 1860s or early 1870s), thus distancing himself from the sanitary science of the early nineteenth century.[19]

Erismann saw hygiene as an experimental science with the analysis of empirical data as its major tool. The experimental character of hygiene meant that for its purposes it also used methods of other related disciplines such as bacteriology, chemistry, physics, or physiology, but even when doing so, hygiene kept its distinct subject matter—"a man in given circumstances from which he may get sick or die." This very word "circumstances [*usloviia*]" distinguished hygiene from physiology that studied the human organism in itself. Hygiene always had to consider the forces located outside the human body and imposed on it by nature and society.[20]

Erismann referred to hygiene also as "sanitary science" and as "the science of public health." For him, sanitation was an "applied public hygiene" that should translate scientific discoveries into practice: "In relation to the external circumstances from which individual or public health may suffer, the task of sanitation [*sanitariia*], on the one hand, is to fight against the general climatic and local circumstances unfavorable to health, on the other hand, to possibly remove from the social organization of life all those moments that disturb the physical well-being of the population."[21]

The emphasis on the unfavorable environment as the locus of disease was a widespread view among nineteenth-century physicians. This scientific belief was grounded in the miasmatic theory that linked disease with filth and the social theory that connected the lack of health to poverty, exhaustion, and deprivation.[22] According to the classical narrative in the history of medicine, at the end of the nineteenth century that broad environmentalist explanation was replaced by the germ theory with its search for specific pathogens, though recent scholarship has shown that the change was not as straightforward, radical, or swift as we once used to think.[23] How did germs and bacteriology fit with the science of hygiene and sanitation in Russia?

Considerable research on the history of bacteriology has shown that it enjoyed early success in the Russian Empire. Some interest in the study of microorganisms was evident already in the 1870s, with a more coherent field emerging in the mid-1880s. By the end of 1887, Russia had no less than eight bacteriological stations, two of them in Moscow. Several of these institutions became important centers of bacteriological research and training. In the early 1900s, the Pirogov Society of Russian Physicians called for the creation of chairs of bacteriology at all medical schools signaling the acknowledgment of bacteriology as a part of medicine's scientific base.[24]

Although there is a certain consensus in historical scholarship about the general acceptance of bacteriology among the Russian medical community by the 1900s and even the convergence of bacteriology and hygiene in the public health work, the earlier decades and the figure of Erismann appear to be more controversial. Some historians describe Erismann as a "thoroughgoing environmentalist," a "formidable skeptic" and opponent of bacteriology. Erismann's biography, in particular, the fact that he was a disciple of Max von Pettenkofer, could give an impression of Erismann as a devoted proponent of Pettenkofer's ideas, including his distrust for bacteriology.[25] It was especially the polemical exchange between Erismann and the bacteriologist and immunologist Ilya Mechnikov, one of the founders of the bacteriological station in Odessa and a future Nobel Prize winner, about the role and significance of bacteriology for Russian medicine and Erismann's defense of hygiene against bacteriologists that received considerable attention from historians.[26] This controversy, however, has obscured the more nuanced relations between hygiene and bacteriology in Moscow in the last two decades of the nineteenth century.

I argue that, by the turn of the 1890s, bacteriological knowledge had already entered both Erismann's teaching at university and the practice of public health in Moscow, but it did not challenge the position of hygiene (and sanitation as its applied version) as the science of public health. Erismann praised the discoveries of Pasteur and Koch and considered bacteriology a "useful tool" for a hygienist.[27] His laboratory at Moscow University included a room for bacteriological research, and the new Institute of Hygiene, built in 1891, had an entire bacteriological department.[28] Already in 1887, Erismann encouraged his students to use bacteriological analysis to search for pathogenic microorganisms, though he was not yet persuaded by the links between discovered bacteria and specific diseases.[29]

By 1891, when the revised edition of his lectures was prepared, Erismann was positive that certain infectious diseases were caused by microorganisms. In his lectures Erismann did not question contagionism as such but, rather, pointed to the insufficient understanding of its mechanisms. The mere fact that pathogenic microbes cause disease did not satisfy him as he wanted to know the details of this causality. The focus on environmental factors reflected his concerns about the limitations of bacteriological knowledge and contemporary understandings of disease etiology. At the end of the nineteenth century, knowledge of microbes and their pathogenic effect was very limited; for many diseases the agents were not yet identified, or if they were, the reliability of each discovery was still under question. Even the verified identification of a disease agent in itself was of little help to an applied scientist such as Erismann. He required a much broader picture that included the mechanisms of its action, transmission, and prevention as well as the understanding of the social context in which disease operated.[30]

Instead of seeing Russian hygiene/sanitation and bacteriology as opponents, I suggest that a more flexible interpretative framework could bring us closer to grasping their relations. In his analysis of the French experience, David Barnes has proposed a concept of "sanitary-bacteriological synthesis," a framework of understanding and combating disease by integrating old concerns about cleanliness and the unwholesome environment with the new knowledge of microbes as the agents of infection.[31] In France, according to Barnes, this integration happened through the language of bacteriology. In Russia, however, it was hygiene that incorporated bacteriology and remained the umbrella science and concept for the public health work on the ground. The study of etiology and epidemiology of infectious diseases was one of many tasks of hygiene, but it did not subsume the entire field. Bacteriological discoveries also did not lead to the radical change of epistemology. Instead, the "new" knowledge of bacteria and the "old" ideas about the risks of decomposing wastes, filth, and bad air existed in synthetic and syncretic combination. The explanations of disease causation that today would be considered mutually exclusive were not seen as such in late imperial Russia and could coexist within the same approach.

Most important, the acceptance of bacteriology did not lead to a reductionist turn in medical knowledge and practice. The incorporation of bacteriological knowledge enhanced the scientific base of public health work in Russia but did not narrow its broad social and environmental focus. Many of the hygienists' professional concerns will sound surprisingly familiar to us in the twenty-first century. They were interested not only in contagious but also in noncommunicable and chronic disease, not only in pathogenic microorganisms but also in other factors that affect human health such as chemical pollution, nutrition, physical activity, occupation, and lifestyle. At the same time, the public health work of many medical practitioners and urban reformers proved it was possible to combine bacteriological methods with the broader socio-environmental vision of improving public health that hygiene was proposing. As I will show later in the book, in Moscow, bacteriological knowledge was integrated into urban sanitary reforms already in the late 1880s (for example, in the projects of the municipal sewage system or the public slaughterhouse), but these new forms coexisted with "traditional" sanitary formats, environmental concerns, and broader social reforms.

The emergence and establishment of hygiene as an academic discipline and the scope of its scientific agenda not only reflected the pressing questions discussed by the Russian society but also developed them and thus contributed to the increased public interest in the problems of disease, pollution, environment, sanitation, and social reform. The preoccupation with external factors that affect human health in a variety of ways meant that hygiene and sanitation brought many social and environmental questions to the light of public attention. Hygienists and physicians discussed disposal and treatment of human and industrial wastes, access to clean drinking water, safety of working conditions and their long-term effects, nutrition, food control, the relation between human and animal health, childcare provisions, health aspects of schooling, physical education, built environment, organization of urban space, and environmental and sanitary legislation.

Many Russian medical practitioners and health reformers shared the values and ideology of the intelligentsia that absorbed the broad ideological spectrum of European liberalism, socialism, and the Russian populist movement of the 1860s. They believed that the privileged minority were called to improve the life of the uneducated and impoverished majority.[32] In the debates on the reorganization of Russian society, however, this ethos often went side by side with the intelligentsia's own claims for greater autonomy and political influence that were denied to them by the autocracy. Medical practitioners saw in their work the potential for overcoming dramatic inequalities and merging with the people and simultaneously stressed their social activism and service to the nation in the quest for the profession's authority and prestige.[33]

For Erismann and for many of his followers, hygiene offered the knowledge and sanitation the tools to promote health and prevent disease. They sent an empowering signal to the Russian medical community: with their knowledge and skills, physicians and medical scientists can significantly improve the life of the Russian population and therefore can and should get involved. Erismann's vision of hygiene was empowering not only for physicians but also for the new zemstvos and municipalities as well. He argued that matters of hygiene could be tackled by the local governments, without the interference of the central state: "The experience of our Western neighbors shows that a too direct interference of the legislation in the organization of public hygiene is absolutely undesirable. The legislation only has to create certain boundaries in the interests of public health that nobody could trespass without being punished. Then it is necessary to remove all the obstacles for the emergence of sanitary organizations in municipalities and zemstvos and to give the local self-government the right to independently create in their localities the institutions that would be responsible for studying the means to improve the sanitary conditions."[34]

Erismann believed that sanitary reforms should be a decentralized process based on "the participation of the local population."[35] Hygiene, with its emphasis on external socio-environmental factors in disease causation and its direct links to social policy, not only provided the necessary scientific grounding for the infrastructural reforms of local authorities but could also enhance the public autonomy from the imperial administration. It is not surprising that this vision resonated well with local governments, including the municipal reformers in Moscow.

Community Medicine, Transnational Municipalism, and the New Politics of Public Health

If the rise of scientific medicine and the establishment of hygiene in the 1860–1880s, the reception of bacteriology, and the emergence of the sanitary-bacteriological synthesis under the umbrella of hygiene in the 1880s–1890s were the key developments in the field of public health as science, the single most important Russian development in the field of public health as politics was the appearance of community medicine, a pioneering program of free rural and urban health carried out by zemstvos and municipalities starting from the 1870s. Before that, medicine in Russia had largely been a state enterprise. The state educated, employed, and supervised the majority of the medical personnel; in fact, until the last third of the nineteenth century, very few positions for medical practitioners could be found outside state institutions. The scope of public health measures and general access to medical care had remained very limited—in the 1860s, the 80 million population of the Russian Empire had only about 10,000 registered physicians.[36]

Even though contemporaries often viewed Russia as an over-bureaucratized state, in the sphere of public health it was in fact under-governed. The functions of medical and sanitary control were scattered, often with overlap and confusion, among several state agencies with little responsibility and executive potential to coordinate any efficient policy on a national level. According to the law, the sphere of civil and veterinary medicine belonged to the Ministry of Internal Affairs, which was responsible for the registration of medical personnel, supervision of state hospitals, education and statistics, anti-epidemic measures, and veterinary control. In addition, the Ministries of Public Instruction, Finance, Transportation, Agriculture and State Domains controlled health-related issues in the structures within their own, sometimes very considerable, jurisdiction. On the local level, medical control belonged to the appointed provincial and city governors. Their administration included provincial medical boards responsible for anti-epidemic and anti-epizootic measures, medical certification, forensic and legal medicine, and the general supervision of public health. On the lowest step in the state system of medical control were the local police who were accountable for the enforcement of the instructions issued by all the authorities involved in sanitary regulation.[37]

The Great Reforms brought about another, potentially more efficient actor in public health. They empowered local communities themselves to become actively involved in the health policy in their regions through the authority of elected self-government institutions and to develop infrastructure to prevent and treat disease. The appearance of new local self-government institutions created numerous alternative employment opportunities for medical professionals. It allowed them to act for and on behalf of the *obshchestvo* (the public), and not the state, while the gap between the two deepened during the conservative turn of the Russian government after the assassination of Alexander II in 1881. Russian obshchestvo saw elected municipalities and zemstvos not as the local state but as parts of the obshchestvo—and so did the municipal or zemstvo activists themselves.[38]

The new concepts of *obshchestvennaya meditsina* and *obshchestvennyi vrach* (roughly translated as "community medicine" and "community physician") entered Russian public discourse in the last third of the nineteenth century. Understanding of these concepts was double-edged—on the one hand, they implied the medical system organized by the local community and functioning according to its demands; on the other hand, they meant the shift from medicine focused on an individual to medicine focused on a community as the object of intervention. As the Kazan professor of pathological anatomy A. V. Petrov described this development in 1873: "after centuries of fruitless service to individuals, medicine and physicians are called to serve the whole society."[39]

Dmitrii Zhbankov, one of the leading figures in community medicine and a member of the populist movement, later described this shift in the following words:

> In his private practice a doctor dealt only with individual patients, not connected to each other, and his task was only to cure the sick, without thinking about what would happen afterward and what was happening around. With the emergence of community medicine, it became necessary to deal with the masses of diseased and the healthy population around them, to get convinced of the tight connection between the sick and the healthy, to see the dependence of diseases and epidemics on all environmental factors. . . . It immediately became clear that it is necessary to fight prejudices, to propagate hygiene, to introduce possible sanitary measures, so to speak, to actively interfere in the people's life.[40]

The community medicine organized by the zemstvos is recognized by historians, both in Russia and abroad, as a major innovation and a social achievement of the post-reform period. Its historic image is intertwined with a certain mythology of devoted idealistic physicians sacrificing their life to serve the poor. Zemstvos opened hospitals and outpatient clinics for the rural population, created training schools for midwives and daycare facilities for children, carried out smallpox vaccination and impressive medical statistical studies. However, the quality of zemstvo health care and sanitary programs varied considerably, and despite abundant literature of the last thirty years on zemstvo medicine and a more critical approach toward the social, ethnic, and gender biases of zemstvo physicians, the necessity of individual zemstvo measures and their adequacy to the specific problems they were supposed to solve in the context of the scientific knowledge of the time (or of today) has rarely been questioned.[41]

Municipal community medicine often appears as a younger and inferior sister of zemstvo medicine that grew under its influence and in its shade.[42] In a big city like Moscow, however, the process was more complex. Clearly, the development of zemstvo medicine in the countryside contributed to the atmosphere of great expectations and a huge interest in public health as politics. The work of the zemstvo in the Moscow province (which surrounded but did not include the city of Moscow) was particularly important, both because of the complex legal, economic, geographical, and personal connections between the municipality and the provincial zemstvo and because the Moscow zemstvo, led by a progressive lawyer Dmitrii Naumov, was one of the pioneers of community medicine and sanitary work of the 1870s.

The head of the sanitary commission of the Moscow zemstvo Evgraf Osipov was convinced that progress in public health was possible only if preventive medicine and curative medicine went hand in hand. He argued that "rational hygiene and enlightened administration can do more for public health than the 'doctor's art.'"[43] To find out where and what kind of preventive interventions were necessary, in the 1870s the Moscow zemstvo initiated large sanitary studies to analyze local variations in disease and their connections to environmental and social factors. One of these studies was the investigation of the sanitary conditions at the factories of the Moscow province, including their impact on the health of workers and the surrounding rural population. In 1878, the zemstvo invited Friedrich Erismann to carry out this study, marking the beginning of his direct involvement in community medicine. The inspection resulted in a multi-volume study that became a ground-breaking work in environmental and occupational health.[44] The specific sanitary initiatives of the Moscow zemstvo, in addition to broader discussions on health and disease in Russia, triggered new political attention to public health and sanitation in Moscow in the 1880s.

However, to fully understand the politics of public health in Moscow, we need to look beyond connections between the zemstvo and municipal work and beyond Russia to transnational interurban networks. In the second half of the nineteenth century, many rapidly urbanizing cities were facing similar challenges similar to those of Moscow: immigration, housing shortage, infectious disease, pollution, and waste disposal. Urban government officials around the world were aware they had to deal with similar problems, and they openly sought and absorbed applied urban knowledge from other places—the process that historians have called "transnational municipalism."[45]

Certainly, interurban exchange had existed long before the nineteenth century, but modern urban growth and bureaucratization combined with improved transportation and communication made this process more explicit and systematic in the decades after 1850. Formal and informal study trips, international congresses and exhibitions, world fairs and architectural competitions, traveling experts, and the press, all created a certain "marketplace" of urban ideas, models, strategies, and technologies, where municipal governments could search for best practices from other cities in the effort to modernize their own. Recent scholarship in urban history and the history of science and technology revises the older interpretation of that modernization as the linear process of borrowing from Western European metropolises. Instead, historians now emphasize the importance of the multidirectional circulation of ideas, people, technologies, and commodities across borders, reconceptualizing the role of colonial and non-Western cities as laboratories of modernity.[46]

Like many other cities, Moscow was integrated into those international urban networks and actively looked for and studied the best practices from far away—and it is important to acknowledge these connections to grasp Moscow's urban politics. From the late 1870s, the official journal of the municipality, *News of the Moscow City Council* (*Izvestiia Moskovskoi Gorodskoi Dumy*) regularly published articles and reports on urban development in other cities, thus introducing and shaping ideas about innovations that Moscow needed and the models it should follow. By the turn of the 1880s, Moscow was inviting foreign experts and was organizing international study trips, while foreign cities became a constant reference point in municipal discussions, including those on public health and sanitation.

Municipal public health in Russia is usually viewed in the context of zemstvo medicine, but in the case of Moscow, it was transnational municipalism that in many ways defined the problematization of specific spheres along with the chosen solutions. In particular, interurban exchange drew the attention of the Moscow public to technological solutions for preventing disease in the 1870s and especially in the 1880s. Whereas the sanitary work of zemstvos was centered on hygiene education, medical statistical research, and sanitary regulations, foreign experience suggested that the way to a healthy city lay in the sphere of sanitary infrastructure, engineering, and technology—an approach that would not appear in zemstvo medicine until the late 1900s.[47] In fact, the identification of a technological solution abroad sometimes preceded the formulation of a respective problem by the municipal government in Moscow and was the trigger, rather than the result, of discussions.

So what had changed between the 1850s and 1880s to make Moscow's "dirt" more visible, the Moscow public more vocal about their city's "unhealthy" condition, and the Moscow authorities willing to take action? The population of the city grew dramatically and the city faced a mounting challenge of accommodating hundreds of thousands of new urbanites, of providing them with food, water, and shelter, and of removing the waste they produced. The post-reform accounts, however, as the quotations at the beginning of this chapter show, described Moscow's pollution and dirt not as a new development but, on the contrary, as a historic, centuries-old, or even "primaeval" practice, as something that had to be overcome despite having been there forever.[48]

The new visibility of Moscow's dirt cannot be explained by the material and demographic change alone—it was also the result of the change of the viewer and the optic. In the second half of the century, after the disaster of the Crimean War and the Great Reforms, earlier visions of social harmony yielded to the loss of public faith in the political regime. They gave way to a deepening sense of Moscow's "backwardness," when proliferating accounts of the city's filth, pollution, and unsanitary conditions reflected not only the physical reality but also the increasing demand for political change and modernization among those who were writing and reading those accounts.[49] The new entrepreneurial and professional elites, empowered by the reforms of education and especially of local self-government, articulated the urgency of civic activism and political involvement. In autocratic Russia, public health, sanitation, and urban "improvement" were precisely the spheres where the elected local government could claim most authority and independence. The newly established science of hygiene, which later incorporated germ theory and profited from bacteriological discoveries, provided scientific grounding to those new politics of public health and linked the quest for clean and sanitary modernity to the local, rather than national, politics and to social reforms and environmental and infrastructural interventions, rather than just interrupting the chain of infection.

CHAPTER 2

Making a Sanitary City

In its pursuit of a sanitary city, Moscow joined many other cities across the globe whose leaders too carried out health reforms, introduced new services and infrastructures in the dynamic period of urban transformation in the last decades of the nineteenth century. The making of a sanitary city was not always a steady and consistent process; it had its conflicts, advances, and retreats. In the period between the mid-1870s until the mid-1880s, dirt, pollution, and disease became a subject of public debate. The reformed Moscow municipality acknowledged its responsibility to tackle these problems, but the realization of that goal proceeded slowly. During Nikolai Alekseev's tenure as mayor from 1885 to 1893, major breakthroughs took place, as key initiatives in public health and sanitary services were planned and Moscow's path of development for the following decades was decided. From the mid-1890s, after the sudden death of Alekseev and after important legal and personnel changes in the increasingly conservative political climate, the sanitary project lost some of its driving energy. However, it was in this period that many of the initiatives conceived under Alekseev were finally implemented and completed.

In this chapter I reveal the breadth of the sanitary project and its evolution from a version close to the zemstvo model, based primarily on the sanitary inspection and normative regulation, to a new model built on complex technological infrastructure. Relations between the imperial administration and the Moscow municipal government in the context of public health and sanitation were also changing. These relations shifted from conflict and mistrust in the 1880s toward a growing acknowledgment of the municipal sanitary project and limited support to it—even if sometimes this support took the form of administrative measures that bypassed existing legal procedures.

Municipalization and the Public Good

In 1881, the Slavophile newspaper *Rus* lamented the state of affairs in the Moscow City Council and the weakness of the mayor: "Our council gathers almost every week, every trifle provokes long and meaningless discussions, the city mayor has no freedom for personal initiative and is constantly shadowed by the council. . . . At the moment the council is only a school for exercising in the public discussion of public questions. It is not useless for those who exercise but is completely useless for the resolution of the questions."[1] Just five years later, the relations between the council and the mayor could hardly raise such concerns. In 1885, the Moscow municipality elected a new mayor, Nikolai Alekseev (1852–1893), whose term in the office became a turning point in various spheres of Moscow's development, including community medicine and the sanitary reform.

When elected mayor, Nikolai Alekseev was only thirty-three years old and would remain the youngest Moscow mayor of the entire post-reform period. Alekseev belonged to the merchant estate and came from one of Moscow's wealthiest entrepreneurial dynasties, which had descended from serfs several generations before that. Despite his young age, Alekseev was a successful entrepreneur in charge of several industrial enterprises and had experience in public politics. He had been a deputy of the Moscow City Council since 1881 and a deputy of the Moscow zemstvo since 1880; he had served as Moscow's sanitary warden and a treasurer at a Red Cross unit.

Although he lacked a formal education (he was homeschooled and did not have a university degree), Alekseev was well connected in Moscow cultural and artistic circles. He was a friend of the pianist Nikolai Rubinstein and the composer Piotr Tchaikovsky. His cousin Konstantin Alekseev— better known under his stage name, Stanislavsky—was a theater director. Curiously, Tchaikovsky lobbied Nikolai Alekseev for the position of the director of the Moscow Conservatory and even offered to publish some of his own musical works under Alekseev's name to improve his chances, but the plan did not work out. It would be urban politics and not music that would make Alekseev famous.[2]

Alekseev was the subject of numerous memoirs, some of them more favorable than others, but both supporters and critics agreed that he was an extraordinarily active mayor committed to urban reform. He took a keen interest in the matters of city infrastructure, sanitation, and community medicine, and during his term in office the municipality established and united under its umbrella a whole network of sanitary and medical institutions that empowered it to independently manage a large sphere of urban public policy. The sanitary reform was gaining momentum in Moscow along with rising tensions between the municipality and the imperial administration, which was concerned with the growing independence and influence of self-government institutions. Opposition had already started in the early 1880s in the mayorship of Boris Chicherin (1881–1883), but the conflict reached its culmination toward the end of the decade when the Governor-General Prince Vladimir Dolgorukov (1810/1865–1891) faced the newly elected Alekseev.

Dolgorukov was a descendent of an old Russian aristocratic family and had made a military career back in the time of Nicholas I. In 1885, when Alekseev was elected Moscow mayor, Dolgorukov was seventy-five years old, and he had been Moscow governor-general for twenty. Dolgorukov was a believer in the ancien régime values and, according to his late-imperial biographer, was a "mouthpiece of the loyal [*vernopoddanicheskikh*] feelings of Muscovites" and a supporter of "the principles of protecting the autocratic power of the tsar, of Orthodoxy and nationality."[3]

Compared to Dolgorukov, Alekseev was the embodiment of the changes taking place in Russian society. He was forty years younger, a merchant by estate and a successful industrialist by profession, independent, energetic, and ambitious. In the words of Dolgorukov, "with his enormous wealth and loud voice," Alekseev "thought he could promote the merchant liberalism in Moscow and in the [city] council."[4] As governor-general, Dolgorukov tried to curb the dangerous "liberalism" and "parliamentarism" in Moscow's municipal institutions, which, in his view, were avoiding the "necessary and beneficial influence of the governmental administration."[5]

The interests of Alekseev and Dolgorukov clashed repeatedly—first in 1886, when Alekseev allowed himself to comment publicly on Russian foreign policy during the emperor's visit to Moscow, then over the issue of the commercial arcades that Dolgorukov had ordered to close, then in 1887 when Dolgorukov opposed municipal involvement in water-pipe construction and the reform of venereal disease control and unsuccessfully tried to remove Alekseev from his post. In 1889, when Alekseev was reelected mayor of Moscow, Dolgorukov appealed to Emperor Alexander III for the nonrecognition of the elections, referring to Alekseev's "lack of respect to authorities, his desire for independence and ignoring the administration."[6]

Alexander III, however, supported Alekseev and confirmed him as mayor of Moscow. In 1893, after Alekseev's sudden death, the emperor allegedly said: "I loved him because he was not about politics, but only about work [*zanimalsia ne politikoi, a tol'ko delom*]."[7] Indeed, Alekseev was not a theoretician—as was, for example, his predecessor in the mayor's office, the liberal lawyer Boris Chicherin. The "merchant liberalism" that Dolgorukov was writing about had little to do with Alekseev's political activity but, rather, revealed Dolgorukov's own understanding of the concept of "liberalism" as a lack of discipline and subordination and a desire for independence—the understanding common among the traditional elites at the time.[8]

Alekseev did not put forward any explicit political program and, in that sense, to use the expression of the emperor, was indeed not "about politics." Yet it was Alekseev's "work" that had important political meaning. The expansion and success of municipal reforms implicitly marked opposition to autocracy through opening an independent public domain. In the words of historian Daniel Brower, "the practice of municipal power, when legitimated by the belief in its value and importance, enhanced the role of municipalities as political entities distinct from the state."[9] This process gave urban elites a sense of autonomy from the autocratic state—although the mode of governance within the municipal domain could itself be quite autocratic.

Some contemporaries noted the despotic character of Alekseev's personality and even "the Alekseevan regime" that he introduced in municipal politics.[10] According to Chicherin, Alekseev "was elected with the majority of votes and from the first moment became a tsar in the council; it was no longer self-government, but rather self-rule on the public grounds," and under him "the assembly was deprived of its independence and became an obedient instrument in the hands of the mayor which is undesirable in public government."[11]

One of Alekseev's strategies to reach his goals was the moderation and manipulation of council discussions. Although projects of urban reforms often provoked heated debates with a variety of competing opinions, the minutes and protocols of the city council meetings reveal that Alekseev fully used his position as chair to skillfully channel the discussion in the direction he wanted. Another strategy employed by Alekseev was to convince the council—and the urban elites in general—with his own behavior. He was a very rich man, and he clearly did not come to municipal politics for financial gains. Even more, for the sake of promoting his cause, he was ready to invest his private funds in municipal endeavors. He refused to receive the mayor's salary, leaving that money in the city budget; he donated huge sums for city initiatives and persuaded other members of the Moscow business elite to do the same. Alekseev covered the expenses of receptions and ceremonies. His funds were used, for example, for the construction of the municipal water-pumping stations and several schools, while his wife donated 300,000 rubles for the construction of the Moscow psychiatric hospital (the hospital bears Alekseev's name to this day).[12] Finally, Alekseev relied on expert knowledge. All major decisions in the matters of sanitation and public health were preceded by experts' discussions in various committees and commissions organized by the municipality. Scientific expertise helped present the activities of the municipality as objectively necessary, rational, and progressive and served to justify the intervention in the lives of Muscovites.

In the 1880s, discussions on sanitation were centered on three major themes—those of backwardness, public good, and municipalization. Yanni Kotsonis has argued that, in imperial Russia, "backwardness" emerged as a self-contained framework of explanation and a kind of ideology in its own right. The educated public agreed that Russia was backward and underdeveloped, and this claim could be used to explain practically any phenomena.[13] Although comparison with the West was not necessarily crucial to that construction of backwardness, it certainly played a role in the imagination of Moscow municipal leaders. All available sources of information—from the personal impressions of the deputies during their trips abroad to the growing sets of statistical data—confirmed the conviction that Moscow was far behind the European cities. Yet, for city reformers, Moscow's perceived backwardness was not irreversible but, rather, was a challenge with which they could cope. As one of the municipal deputies exclaimed in 1885, "in the five years that I have been a deputy, council members always discuss how the municipality should make a step in urban reforms so that Moscow resembles a European city."[14]

Overcoming that "backwardness" and becoming "more like a European city" was presented as the "public good." The "public good" of making a sanitary city in Moscow rhetorically embraced the formula of "serving the people," which had dominated the circles of the Russian intelligentsia since the 1860s. At the same time, this framing also referred to the patriotic and Slavophile values of the Moscow merchants who wanted to place their businesses at the service of the nation.[15]

The practical mechanism of overcoming "backwardness" and achieving "public good" was seen in municipalization that in the 1880s became the basis for the city's policy in the sphere of sanitation. Both city council deputies and the invited experts perceived the municipal government as the key guardian of public health.[16] The process of municipalization, the expansion of the city activities and jurisdiction, was by no means peculiarly Russian— it was a phenomenon known to the whole Western world in the nineteenth century. In many European and American cities, municipalization was a response to the failure of market forces to cope with the complex problems of urban life.[17] In Russia, however, municipalization had a particular symbolic value and was incorporated in some specifically Russian discussions. The claim of being the mouthpiece of certain social groups or of "the people" was typical for all participants of the political debate. Yet in the autocratic Russian state, only municipalities and zemstvos, which until the early twentieth century remained the highest elected governing bodies, could justify this right in legal terms even though, in reality, their electorate was very limited and not at all representative of "the people." The Moscow municipal elites portrayed themselves as the only legitimate delegates of the city population, as acting on behalf and for the benefit of the urban community.[18]

Not only did Moscow municipal leaders see their task as promoting sanitation, but they claimed a monopoly in this sphere, against both the administrative bodies and the private service providers. When the opponent in the debate was the imperial administration, it was presented as inefficient, arbitrary, and too coercive. When it was the private business the municipality was arguing against, the entrepreneurs were described as "undisciplined," "ignorant," and driven by profit at the expense of compliance with sanitary regulations and public good.[19] According to this rhetoric, similar to the program that Erismann drew in the 1870s, only the elected municipal institutions were responsible, competent, and efficient enough to create a sanitary city.

Municipal Sanitary Reforms in the 1880s and 1890s

Until the mid-1880s, although the questions of making Moscow healthier were debated in the city council, actual sanitary measures were small in scope and were focused in two main directions: sanitary inspection and normative regulation through sanitary decrees. The main concern of sanitary decrees was waste, highlighting the role that this topic played in the image of a healthy city. The very first sanitary decree, issued in 1875, regulated waste disposal and cesspool cleaning; the next one, issued in 1879, prescribed the proper upkeep of cesspools and backyards to prevent the pollution of rivers and groundwaters.[20] The goal of the sanitary inspection was to control the implementation of sanitary decrees as well as to detect and isolate the sick. The sanitary inspection was set up in 1878 as a temporary response to the smallpox and typhus epidemics, but then it kept on being reintroduced the following years.[21] Municipal involvement in health care had also been very limited. The city had only two municipal hospitals: the Shcherbatov Hospital (1866) and St. Vladimir Children's Hospital (1876).

FIGURE 2.1. Alekseiev Municipal Psychiatric Hospital. From *Al'bom zdanii*. National Electronic Library (Russia).

Alekseev, during his term in office, saw a dramatic expansion of municipal health care. In 1887, six of Moscow's hospitals that had previously belonged to the State Welfare Department or to the Department of Empress Maria were moved to the management of the city. Together with a newly opened municipal clinic for chronic diseases, this meant that the city now had nine public hospitals. In 1889, Alekseev initiated and personally raised funds for the construction of a new psychiatric hospital, which would be completed only in 1894, after his death, becoming the tenth municipal hospital (see figure 2.1).[22] Another important health-care innovation of the Alekseevan government was the establishment of so-called *ambulatoriia*—municipal outpatient clinics. The first two outpatient clinics opened in 1886; in 1895 there were already seven of them, serving almost 200,000 individual patients annually (see figures 2.2 and 2.3). All medical services, prescriptions, and medications in municipal outpatient clinics were provided free of charge.[23]

It was, however, in the sphere of sanitation that Alekseevan reforms reached their true breadth. In the decade between 1884 and 1893, expenses for street cleaning doubled, those for the maintenance of public toilets and sewage dumps increased more than sixfold, and for sanitary measures more than tenfold.[24] In 1884, the city established the Temporary Executive

FIGURE 2.2. Rogozhskaia ambulatoriia. From *Al'bom zdanii*. National Electronic Library (Russia).

FIGURE 2.3. Rogozhskaia ambulatoriia. Waiting room. From Uspenskii, *Moskva*. National Electronic Library (Russia).

Sanitary Commission, which, despite its name, became the permanent municipal institution of sanitary control. Sanitary physicians who formed the core of this sanitary commission had the right to enter private houses, shops, factories, and institutions; from 1886, they took over epidemiological control and medical statistics that had previously belonged to the police. Each of the sanitary physicians was in charge of one of Moscow's twenty sanitary districts, where they were responsible for investigating both environmental and social conditions such as water and soil pollution, waste disposal, housing and working arrangements, and the relation of those conditions to morbidity and mortality.[25]

In case of an outbreak of an infectious disease, the municipal disinfection brigade, established in 1889, cleaned or sprayed dwellings and institutional premises with mercuric chloride, carbolic acid, or formalin—internationally used antiseptic and biocide compounds that were also highly toxic. Clothes, textiles, and furniture were either burnt (usually with some compensation to the owner) or sent to the municipal disinfection chamber. All disinfection procedures, although free of charge, were intrusive, materially damaging, and possibly quite frightening for an average urban dweller, as they meant a crowd of disinfection workers in protective gear and with bulky devices entering and wrecking their home and taking away their belongings. Interestingly, however, the municipality reported that resistance to disinfection was rare.[26]

The scope of municipal activity that fell under the umbrella of "sanitary measures" changed. If previously the focus had been almost exclusively on the problem of waste, in the Alekseevan period medical control was extended to other spheres of urban life, including education, food supply, and human-animal relations, which were all now interpreted in medical terms and integrated into the sanitary project. This extension happened both through normative regulation, with new sanitary decrees about food production and the keeping and slaughtering of animals, and through the establishment of new sanitary offices such as the school sanitary inspection in 1889 (discussed in chapter 10) and the trade and market sanitary inspection in 1890. The importance of animals in the vision of the sanitary city was emphasized through the creation of the municipal veterinary organization in 1889–1891, which was responsible for inspecting and treating urban livestock and enforcing the regulations against zoonotic diseases (see chapter 6). Veterinary inspection was described and classified in the city budget as a "sanitary measure" and was listed and financed together with the work of sanitary physicians, street cleaning, and waste removal.

Another new measure, not intuitively sanitary but classified as such in municipal documents, was the control of venereal disease and prostitution.

In respect to prostitution, Russia followed the French model of state regulation: women engaged in commercial sex had to register with the police and undergo regular gynecological examinations at the Medical-Police Committee (Vrachebno-politseiskii komitet). Their passports were taken away and replaced by so-called yellow tickets, a prostitute's identification that immediately revealed a woman's occupation in any situation when she was asked to present her documents, such as when renting a room or finding employment, and was highly stigmatizing. The decision as to who should be considered a prostitute was taken by the police quite arbitrarily and was very difficult to revoke. Many physicians believed that the system was inhumane and also utterly inefficient in preventing the spread of syphilis because it made women avoid medical examinations.[27]

This view appealed to Alekseev, who was convinced that public health measures should belong to the city and not to the police. As a result, Moscow pioneered a new system of municipal venereal disease control. In 1889, the Medical-Police Committee was dismantled, the yellow tickets were abolished and the passports returned to their owners. Instead, women engaged in commercial sex received special sanitary cards, with a photograph but without a name so as to preserve their anonymity. These sanitary cards were valid only with a recent "visa" from a gynecologist and could be checked by brothel owners, by the police, or by vigilant clients. To get such a visa, a woman was supposed to go for a free weekly medical examination at the so-called Sanitary Bureau (sanitarnoe biuro) with a special women's outpatient clinic. This examination was in principle voluntary, and the Sanitary Bureau tried to make the process easier for women (and therefore more efficient in venereal disease prevention) by protecting their anonymity and by employing female personnel, despite the resistance of the imperial administration. Two out of seven physicians and all medical assistants at that outpatient clinic were women.[28]

Many of these sanitary reforms were developed with the involvement of Friedrich Erismann who was the usual member of several expert commissions whose members advised the Moscow municipality. Erismann was even elected the head of the committee responsible for the new policy against venereal disease, although venereology was far away from his specialization and although several syphilologists were nominated for this position.[29] The fact that Erismann was involved in decision-making in several municipal bodies favored the dissemination of his vision of hygiene and contributed to the advancement of his own career. When, in 1891, the city established a sanitary station for the laboratory examination of water, food, and other commodities, Erismann was predictably appointed its director, which endorsed his authority as the key expert in sanitary science.[30]

The system of community medicine and sanitary control that had evolved out of these measures by the end of Alekseev's mayorship had a collegial character and relied on wide circles of medical professionals, who gained an unprecedented power to intervene in and regulate the lives of Muscovites and to turn their social and scientific ideas into urban policy. In a situation of institutional immaturity and low competence in matters of public health, this collegiality provided indispensable advisory support to the municipality and allowed for regular and open professional discussions, where a variety of opinions and firsthand observations made up for lack of experience and necessary training. As a result, public health policy in Moscow developed in a peculiar technocratic system, where key decisions were shaped by invited or appointed experts who were directly involved in the work on the ground but who in many cases did not themselves have voting rights in the city unless they owned a property or a business. These experts took part in policymaking through numerous commissions that either purposefully designed specific reforms or managed the daily work of the sanitary organization. Participation in municipal advisory bodies and consulting local authorities was expected even from the ordinary medical and sanitary personnel, including sanitary physicians, veterinarians, market inspectors, regular physicians of the city hospitals and outpatient clinics; their vision of what the city, society, and environment should look like had a direct impact on urban reforms.[31]

The medical and sanitary organization that appeared under Alekseev bore many striking similarities with the zemstvo community medicine and, specifically, with the endeavors of the Moscow zemstvo. This similarity is hardly surprising considering that Nikolai Alekseev, Friedrich Erismann, and several other members of the municipal sanitary organization had been involved in the Moscow zemstvo before. Both the rural and the urban systems shared the focus on sanitary regulations, sanitary inspection, and the territorial system of health care that was organized, owned, and run by the public government and that provided services free of charge. The municipal veterinary institutions and the municipal interest in animal disease developed on the background of the professionalization and institutionalization of veterinary medicine in the zemstvo Russia and clearly under its influence. Again, there were direct personal connections: veterinary measures in Moscow were devised by Valentin Nagorskii, the chief veterinarian of the Moscow provincial zemstvo. It is plausible that the attention to animal disease was in itself an influence of the zemstvo sanitary model from the countryside, where human and animal populations lived in close proximity and where epizootics and zoonosis posed a far greater threat than in Moscow.

However, this period also marked a major divergence from the zemstvo sanitary model. Alekseev pushed the municipality toward a new approach based on technological solutions to sanitary problems. Technological solutions required not only the commitment of the city government to establish and finance expensive infrastructural projects but also huge investments that the municipality had to procure through obligation loans. The first of such reforms was the municipalization and medicalization of slaughter and the establishment of the public abattoir in 1886–1888 (see chapter 7). The second was the reconstruction and expansion of the water-pipe system in 1890–1892, which allowed the supply of water to increase from 6,000 to 16,000 cubic meters daily.[32] The most important and expensive project of sanitary technology was the sewerage system, which would be launched only in 1898, five years after Alekseev's death (see Chapters 3 and 4).

Unlike sanitary decrees that targeted primarily individuals and private businesses, these new technological projects dealt with the city and its resources as a public domain. They constituted a new type of public property that was supposed to change the dynamics of public health and that was managed by the public government in the name of the public good.[33] They were an attempt to construct a sanitary city not through normative regulation but through changing the material circumstances of urban life, and the impact of this change—in theory—would be felt by all city residents. At the same time, this approach turned the municipal government into a kind of entrepreneur that owned and managed complex enterprises and could eventually extract profit from them, even though this was not Alekseev's original intention.

The Sanitary City and the Imperial Administration around 1900

In the 1890s, several key developments reshaped sanitary reform in Moscow. The major structural change was the new Municipal Statute of 1892, which limited the independence of municipal bodies and increased the control of imperial administration over them. Another innovation was the change of the electoral system. The tax-payment qualification was replaced by a new one based on property ownership, thus reducing the already small number of voters in Moscow from 23,000 to 6,000. Historians have pointed out, however, that in most Russian cities, and Moscow was no exception, only a minority of those entitled to vote were actually coming to the voting stations, and the mass abstention, especially among the poorer taxpayers, had effectively reduced the electorate even before the new legislation.[34] It is still remarkable that, in a rapidly growing city whose population in the 1890s was approaching one million, less than 1 percent of the urbanites had the right to vote. The new statute helped preserve existing trends in the Moscow City Council, prevented the involvement of wider groups in electoral politics in the following decades, and strengthened the positions of merchants and honorary citizens who had already formed the core of the council throughout the 1880s. In the early twentieth century, municipal government was increasingly becoming a task of the educated and socially privileged. If, in the early 1880s, about 33 percent and, in Alekseevan Moscow, about 15 percent of the deputies came from lower social groups such as peasants, workers, or petty bourgeois, by the eve of the First World War, this proportion had dropped to just 8 percent. At the same time, the part of deputies with a higher education grew from 20 percent in 1889 to 47 percent on the eve of the war.[35]

There were also some important personal changes. In March 1893, Nikolai Alekseev was murdered by an insane person right in the new building of the Moscow City Council. After the death of Alekseev, the Moscow municipality never produced such a charismatic leader. The mayors who succeeded him—Konstantin Rukavishnikov (1893–1897) and Prince Vladimir Golitsyn (1897–1905)—governed Moscow in a much calmer manner and avoided open confrontations with the administration until the revolutionary crisis of 1904–1905.[36]

Another loss for Moscow's health reformers was the departure of Friedrich Erismann. In 1895, he supported the riots of the university students and petitioned the imperial administration against the punishment of protestors and the interference of police in university affairs. His petition provoked a harsh reaction within the Ministry of Public Instruction, which reprimanded Erismann, making him quit Moscow University and go back to Switzerland in 1896. This was, however, not the end of Erismann's career—he joined the Swiss Social-Democratic Party and was elected a member of the city council in Zurich, where he worked until his death in 1915. Interestingly, as Pavel Ratmanov has recently shown, while in Zurich Erismann met young Henry E. Siegerist, who went to school with Erismann's children and knew him well—a remarkable connection between a leader of Russian community medicine and a person who would later become one of the most vocal international advocates of socialized medicine and Soviet public health.[37]

The 1890s and 1900s were a period of growing involvement of the imperial administration in the Moscow sanitary project, facilitated both by personal changes and the legislative innovations of the counter-reform era. In 1891, the elderly Governor-General Vladimir Dolgorukov was replaced by Grand Duke Sergei Alexandrovich Romanov (1857–1905), the uncle and brother-in-law of Emperor Nicholas II. Grand Duke Sergei was a political hard-liner, notorious for his conservative views and repressions of radical elements. The same can be said about his main adviser, the chief of the Moscow police Dmitrii Trepov (1855–1906).

Trepov assumed his post in 1896, replacing Aleksandr Vlasovskii, who had been fired in connection to the Khodynka tragedy when more than a thousand people were killed in a stampede at the coronation festivities. Trepov had a remarkable influence over the governor. Minister of Finance Sergei Witte called Trepov the "colleague-boss of Grand Duke Sergei."[38] According to the account of the Moscow City Council deputy Nikolai Vishniakov, Trepov "rules over the city affairs, things happen as he wants them, and the Grand Duke only follows him."[39] Both Trepov and Grand Duke Sergei were particularly sensitive to the questions of unrest and revolution. In 1878, the governor of Saint Petersburg Fyodor Trepov, the father of Dmitrii Trepov, was wounded by a member of a revolutionary organization, Vera Zasulich. Three years later Emperor Alexander II, the father of Grand Duke Sergei, was killed by terrorists. Revolutionaries were not only their political opponents but also a real threat to their lives. Dmitrii Trepov would survive several murder attempts, become the deputy interior minister and the suppressor of the 1905 Revolution, and die his natural death from a heart failure in 1906. Grand Duke Sergei would be killed in 1905 by a bomb thrown by Ivan Kaliiaev, a member of the Socialist Revolutionary Party.

Trepov was notorious for his army manners and managerial style, stubborn authoritative character, love for military discipline, as well as his commitment to fighting the revolutionary movement.[40] What is less known, however, is Trepov's strong interest in questions of sanitation, public health, and pollution. These concerns fitted well with his obsession with order and his fear of social unrest, which, as the 1892 devastating cholera epidemic outbreak had so powerfully demonstrated, could be caused not only by revolutionaries but also by epidemics.

The 1892 cholera epidemic combined with the famine of 1891 was a turning point in the history of Russian public health. The empire was hit by the epidemic when the threat of cholera had declined in most European societies, and when the etiology of the disease and the mechanisms of its prevention were by and large understood. The fifth cholera pandemic spared Western Europe (with the notable exceptions of Hamburg and Italy), while in the Russian Empire the outbreak claimed more than 200,000 lives.[41] The epidemic triggered reforms in the fields of public health and welfare across the empire and contributed to the professionalization of medicine. Combined with the famine, the epidemic also discredited the autocratic regime in the eyes of the educated public, giving way to alienation and conflict between the state and society in Russia.[42]

In Moscow, however, the cholera epidemic led to cooperation between the imperial administration and the municipal government in face of the disease and established a pattern of administrative persecution of those who violated sanitary norms, a persecution that would persist in following decades. In 1892, in response to the cholera threat, Governor-General Grand Duke Sergei, using the power of the Decree on Measures for the Preservation of State Order and Social Stability (Polozhenie o merakh k okhraneniiu gosudarstvennogo poriadka i obshchestvennogo spokoistviia), introduced a new stage in urban sanitary control. Despite the achievements of the municipality in the sphere of community medicine, many sanitary regulations concerning private property remained ineffective because the city had little power to control their implementation. When the violation was recorded, the case was sent to the court, which could only impose a minor fine; in certain cases, paying those fines was easier and cheaper for the violator than complying with the municipal norms.[43] According to the new order, which remained in force until 1894, all those who did not comply with the sanitary rules of the city council were subject to a huge fine of 500 rubles (instead of maximum 50 on the sentence of the justice court) or even three months of imprisonment. More important, the sanitary violators were persecuted not according to the standard lengthy court procedure but simply by the arbitrary decision of the governor-general.[44]

The Decree on Measures for the Preservation of State Order and Social Stability was promulgated by Alexander III in 1881 as a part of his counterrevolutionary attack. It allowed a regime of reinforced security and stipulated that the governors of the regions where state order was under threat had the right to issue special orders, noncompliance with which would result in a 500-ruble fine or a three-month prison sentence. This decree, however, did not concern the questions of pollution; it was conceived as a tool of the security police and aimed primarily to counter political threats.[45] In 1892, this decree was applied to strengthen and support the municipality—although, just several years before, the imperial administration had viewed the city council as the dangerous hotbed of "merchant liberalism" and opposed the expansion of municipal services. In the moment of emergency, the administration and the municipality acted in tandem.

Although this was a special cholera-related measure, at the end of the 1890s it was reintroduced because of the joint efforts of Dmitrii Trepov and Grand Duke Sergei. In 1896, citing several notable cases of sanitary negligence presented by the municipal sanitary commission, Grand Duke Sergei again requested the extraordinary right to punish sanitary violations with his own power, avoiding the legal court procedure.[46] However, by that time, the threat of cholera had diminished, and the sanitary violations could not be easily portrayed as a danger to the state. After a year of intensive lobbying, Sergei succeeded in obtaining the right to punish violators out of court for the following five years as a special permission of Emperor Nicholas II.[47] According to Trepov's reports, the introduced measures seemed to work. If, in 1897, 63 percent of the inspected houses did not comply with the municipal requirements, by 1902 that proportion had dropped to 17 percent.[48] In 1903, the special powers of the governor-general in administrative persecution of sanitary violators were prolonged for another five years.[49]

The new system of persecuting sanitary violations that emerged in Moscow demonstrates that, by the turn of the twentieth century, the imperial administration had transformed into an occasional ally of the Moscow municipality in questions of public health. Plausibly, given the public attention received by health topics in Russia after the cholera epidemic of 1892, the support of the local administration for municipal initiatives can be interpreted as an attempt to receive credit for sanitary measures originally developed by the self-government institutions in the format of community medicine. On the other hand, this new cooperation between city government and imperial administration in matters of sanitation reveals the limits of the municipal sanitary project and municipal rule in Russia more generally. Despite all its impressive endeavors and expertise and despite legitimation through public representation, municipal authority still yielded in importance to the administrative resource. Actors on all levels of the government admitted that the power of the municipality to enforce sanitary regulations was limited and that court procedures were inefficient. The resolution of this problem, however, did not follow the path of legally expanding the authority of the municipality or making the courts more efficient but, instead, introduced extraordinary restrictive measures based on bureaucratic discretion that bypassed the existing law.[50]

The Sanitary City on the Eve of the First World War

The first decade of the twentieth century saw a major crisis connected to the first Russian revolution of 1905, which started with Bloody Sunday in January in Saint Petersburg and reached its even bloodier culmination with the December Uprising in Moscow.[51] The revolution resulted in the creation of the first Russian parliament—the State Duma. In cities, electoral rights were now given to all those who owned any real estate property, who rented separate apartments, ran businesses, or were employed by the state, municipal, and zemstvo organizations or by the railways. For the first time (male) civil servants, professionals, clerks, and qualified workers living in rented apartments became entitled to vote. These new rights, however,

FIGURE 2.4. Soldatenkov Municipal Hospital. Surgery department. From *Al'bom zdanii*. National Electronic Library (Russia).

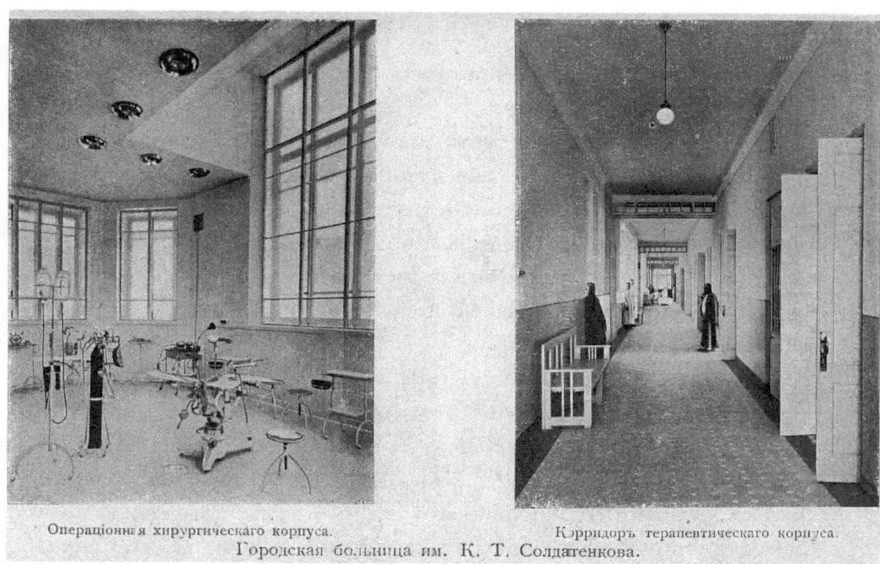

FIGURE 2.5. Soldatenkov Municipal Hospital. Operating room at the surgery department and the corridor at the therapy department. From *Al'bom zdanii*. National Electronic Library (Russia).

FIGURE 2.6. Morozov Municipal Children's Hospital. From *Al'bom zdanii*. National Electronic Library (Russia).

extended only to the State Duma elections; city elections continued to be based on a much more conservative voting system.⁵²

Still, the revolution gave a new impulse to social policy in the city. It left Moscow elites in a state of shock and led to the desperate search for the means to pacify and prevent unrest. Consequently, the post-1905 years saw a surge of municipal activity in health care and schooling, the expansion of sanitary services to outer districts, the first attempts at housing reforms, as well as significant improvement in working conditions at the municipal enterprises.

An important tendency of the early twentieth century was the bureaucratization of the medical and sanitary organization, with the appearance of new offices and officials whose task it was to manage that rapidly growing branch. Its collegial and democratic character was, however, preserved. Various groups of regular municipal health professionals—physicians at hospitals and outpatient clinics, sanitary doctors, school doctors, obstetricians, and psychiatrists—continued to influence city politics through their respective advisory committees and by electing members to the municipal medical council, which discussed and evaluated all the health-related policies of the municipality and, importantly, voted on the candidates for the vacant positions in the city's medical and sanitary institutions.[53]

Another tendency was the expansion and growing specialization of the municipal health-care services. Of the nineteen city hospitals in 1912, ten had a specialized focus such as ophthalmology, pediatrics, psychiatry, gynecology and neonatology, venereal or chronic diseases. The treatment in municipal hospitals, except for psychiatric and chronic patients, was free of charge for all those who had paid the small hospital tax (1.25 rubles per year). Outpatient health care also became specialized—by 1910 Moscow had 29 outpatient clinics, with separate clinics for pediatrics, ophthalmology, otolaryngology, and dentistry (see figures 2.4, 2.5, and 2.6.) Together, these clinics were serving more than one million individual patients per year, and all their services continued to be provided free of charge.[54]

A third tendency was the growth and increasing importance of technological infrastructure such as the abattoir, the sewage system, and the water pipe. In addition to the water-pipe system reconstructed under Alekseev, the city built another large system with filtered water from the Moskva River, and the daily supply of piped water increased six times compared to the early 1890s (or fifteen times compared to the 1870s). In the last years before the First World War, the maintenance and operation of these three sanitary enterprises accounted for three-quarters of municipal sanitary expenses and one-third of all city expenses related to medicine, health care, and sanitation in general. The expansion of technological infrastructure also resulted in another new aspect of the Moscow sanitary project—it started to make a profit. First, the water-supply system and the public abattoir and then also the sewerage system began to generate enough gains to cover operational expenses and loan payments. The annual net profit from these three enterprises grew from 240,000 rubles in 1901 to 830,000 rubles in 1910 and to just over one million in 1911, a sum that was sufficient to cover almost all other municipal sanitary expenses.[55] Preventive sanitary measures—which back in the 1880s were seen as a necessary expenditure that the government must incur for the sake of public good, for saving lives and fighting disease—turned into a source of municipal revenues, which, combined with large private donations, helped finance the expanding social and public health services in Moscow on the eve of the First World War.

The narrative of sanitary reforms in Moscow so far has been one of success. When we look at the general development of the sanitary project between the 1880s and 1910s, it is easy to focus on expansion: the impressive diversification of the public health and sanitary services; the increased number of hospitals, outpatient clinics, physicians, and patients served; the growth of the city's expenditure on public health and of the profits from the municipal enterprises. However, such a focus masks the many limitations and inequalities of the sanitary project and cannot explain why the mortality rates in Moscow remained so high. Sanitary reforms inevitably remained a selective endeavor; they concentrated on some urban problems and neglected others. To understand what that quest for a clean and healthy modernity meant for the population and the environment both inside and outside of Moscow, why some parts of urban life became problematized, who benefited from that quest and who was left out, it is now important to consider specific sanitary reforms individually.

PART II

Water, Waste, and Technologies of Sanitation

CHAPTER 3

The Sewage System

European Symbol, American Design

During the nineteenth century, many European and American cities created new infrastructures to manage water and waste. Cities were paved and asphalted, preventing rainwater from permeating into the soil, and networks of gutters conveyed precipitation to the surface waters. The velocity and locations of water flows changed; rivers were regulated to avoid flooding and to facilitate navigation. Large reservoirs and aqueduct systems were constructed to capture, filter, and distribute rainwater in the cities; more groundwater was piped in to satisfy the needs of the population and the economy. The accumulation of human sludge in cesspools and privy vaults was replaced by networks of drains and sewers and sometimes also by the complex systems of waste treatment.[1]

Throughout most of the nineteenth century, it was the miasmatic theory that provided the background and rationale for developing technologies of sanitation. The miasmatic theory was based on the assumption that disease originates from decomposing organic wastes and is spread through their emanation. The depositories of "filth," including cesspools, dumps, or waterways, were perceived as sources of contamination. Environmental sanitation and the prompt removal of decaying organic materials to prevent the spread of their rotting smells were thus seen as essential for fighting epidemics. In the second half of the nineteenth century, belief in the

TABLE 3.1. **Urban Water and Sanitary Infrastructure in the Russian Empire in 1910**

Region	Total Number of Cities	Cities with Water Pipes	Cities with Sewerage
European Russia (50 provinces)	862	168	40
Poland	128	10	9
Caucasus	137	31	14
Siberia (including Far East)	52	5	2
Central Asia	52	5	0
Total	**1,231**	**219**	**65**

Source: Statisticheskii ezhegodnik Rossii za 1915 g.

connection between disease and filth and in the health value of flushing away household wastes stimulated many cities across Europe and America to build or reconstruct their sewers—Paris in the 1850s, Chicago in 1859, London in 1859–1867, Vienna in 1861, Hamburg in 1862–1869, Frankfurt in 1867, Berlin in 1873–1893. In the last decades of the nineteenth century, the "bacteriological revolution" replaced filth, miasma, and foul odors with germs as the causes of disease, but the commitment to develop wastewater infrastructure persisted.[2]

The cities of the Russian Empire also joined that international trend of building sanitary infrastructure. The country's first complex urban sewage system was built in Odessa in 1877. Warsaw, the empire's third largest city, constructed new waterworks and a sewage system in 1883–1886. Moscow rebuilt and expanded its water-supply network in 1889–1903 and launched the sewage system in 1898. By the eve of the First World War, sewage systems existed in sixty-five Russian cities (see table 3.1).[3] As the imperial capital, Saint Petersburg, did not have modern sewerage (it was built only in the Soviet period), the Moscow sewage system remained the biggest project in waste removal infrastructure in the Russian Empire.[4]

How and why did the Moscow sewage system come into being? It is easy to imagine the sewage system as the logical answer to the challenges of urbanization, but the example of Saint Petersburg, where similar demographic pressures did not result in such infrastructural solutions, indicate

The Sewage System

FIGURE 3.1. The Moskva River. View from the Ust'inskii Bridge. From *Moskva. Vidy nekotorykh gorodskikh,* appendix 1 (1888). National Electronic Library (Russia).

that this relationship was more complex and problematic. The first blueprint for the sewage system in Moscow was developed in 1874. The final and very different design, approved in 1890 and commissioned eight years later, was modeled not on Paris or Berlin but on the sewage system in Memphis, Tennessee. Once we acknowledge that conditions do not determine responses, a number of fascinating questions come up. Why did Moscow's authorities become interested in waste removal at that specific moment? Which political, cultural, and environmental factors influenced the final shape of the project? How did Moscow's aspirations to be like a European metropolis lead it to adopting a sewerage design from a small American city?

The two decades of discussion and construction of the Moscow sewage system coincided with a considerable shift in the understanding of disease, connected to the bacteriological discoveries and the rise of germ theory and laboratory medicine.[5] Similar to many sanitary projects of the period, the construction of the sewage system in Moscow was informed by changing contemporary science and the new preoccupation with germs. However, the decision about what kind of scientific advice and technical expertise to consider was highly political. That decision depended not so much on who offered the "best science" but on how well their involvement and proposed solutions would fit with the values and objectives of politicians. As Moscow politics changed, moving toward municipalization and more ambitious, active, and interventionist local government, so did the project of the sewage system.

River Environments, Pollution, and Public Health

As with many cities, rivers played a crucial role in the history of Moscow. The city emerged on the banks of and also derived its name from the Moskva River (see figure 3.1), which rises in the Smolensk-Moscow Upland and flows 500 kilometers to the Oka River, a tributary of the Volga. Up until the construction of railways, the Moskva River remained the most important means of transportation to and from the city. Reliable commercial navigation was, however, hindered by the river's general shallowness and the instability of its current—in the 1870s, to cover the 180 kilometers between Moscow and Kolomna, the town at the mouth of the river, ships needed two weeks in high water and up to two months when the water was low.[6]

The Moskva of the late imperial period was very different from the river that we know today: the completion of the Moscow Canal in 1932–1937, one of the Gulag construction projects, diverted some water from the upper Volga into the Moskva River and stabilized its flow. Throughout the nineteenth century, despite projects in river regulation, spring floods connected to the rapid snowmelt remained the norm, with a particularly disastrous flood occurring in 1879. In 1882, a city engineer described the Moskva River in the following words: "a weak current with a minimal vertical drop, with a speed below 0.15 m/s and a streamflow of 30 m3/s when entering the city and 50 m3/s when leaving it. This is how the Moskva looks eleven months a year; but then spring comes, ice breaks, and the river becomes unrecognizable: its waters rise rapidly, sometimes 8 m above its normal level, its speed increases to 2 or even 3 m/s, the volume reaches 1600 m3/s and sometimes 2600 m3/s; the river spills and floods 1/7 of the city territory, causing numerous calamities."[7]

In the nineteenth century, the Moskva River and its inflows, with the forty-eight-kilometer Yauza being the biggest one (see figure 3.2), remained important sources of water for Muscovites. Although, since 1804, the city had a water pipe that brought water from the artesian wells in the village of Mytishchi, northeast of Moscow, the supply was not sufficient to cover the needs of urban dwellers. In 1879, the pipes provided 7,800 cubic meters of water daily, which was far below the needs of the city with a population of 700,000. The reports of municipal sanitary physicians confirm that the river and well water, despite its poor quality, were almost universally used

The Sewage System

FIGURE 3.2. The Yauza. View from the Vysokoiauzskii Bridge. *Moskva. Vidy nekotorykh gorodskikh*, appendix 1 (1888). National Electronic Library (Russia).

not only for laundry and bathhouses but also for drinking and cooking. Although additional smaller water pipes were built, by the 1870s the deficit of drinking water became a pressing issue in the rapidly growing city.[8]

The network of the Moskva River and its tributaries also served as an excretory system for the city. The vital metabolic function of Moscow's five rivers and twenty-two streams was to deliver the city from the rain, snow, and wastewater—and solid wastes to some extent as well. This was the task of many urban rivers, but Moscow's situation was complicated by the very moderate speed, depth, and streamflow of the Moskva River. According to the hydraulic estimations of V. I. Astrakov in 1878, the streamflow of the Moskva River was only half of the Seine in Paris, four times less than the Tiber in Rome or of the Rhone in Lyon, and seventy times less than the Neva in Saint Petersburg.[9] Although Moscow's waterways for centuries were used as deposit for the urban wastes, in the 1870s their pollution came into public debate, and soon it was acknowledged that the problem required a solution in the form of a sewage system).[10]

The objective reality behind this change in perception of urban metabolism was the city's demographic and industrial growth. In the two decades after the abolition of serfdom, the city's population doubled. Furthermore, by the end of the 1870s, Moscow had almost 500 factories with 120,000 workers that intensively used urban water resources.[11] The disposal of the waste left by the increasing population of Moscow indeed posed a challenge. According to contemporary estimations, each person produced 11.5 tons of waste annually, and even after evaporation, this left almost 8 tons to be disposed of.[12] Moscow relied on the cesspool system to reach this goal. Legally, each household was supposed to accumulate its wastes in a cesspool, the contents of which had to be regularly removed to the city dumps by special cesspool cleaning carts to prevent noxious odors and the pollution of water and soil.[13]

The practice was, as often happens, quite different. Moscow landlords delayed or avoided calling the cesspool cleaning carts and invented alternative (and cheaper) means to get rid of wastes. Although the municipal and governmental decrees prescribed that natural streams should be kept clean and strictly forbade dumping human excrement in the city rain drains, numerous accounts testify that those regulations were frequently ignored. As most of the small rivers and brooks went through private estates, their owners used the natural streams to drain the refuse, either directly or through underground pipes. The simplest solution, however, was to throw the sludge into the street ditches and gutters on rainy days and, especially, nights when the streams of stormwater carried it away to the river.[14]

In fact, even the regular calling of the cesspool cleaning carts did not ensure that the sludge would not end up in rivers. Thus, in 1874, the residents of Moscow's suburb Shiriaievo Pole petitioned the city mayor to take measures against the pollution of the Yauza River that "produces such a stench that it has become dangerous for our health to live in our houses."[15] The river water, according to them, was deteriorating because the workers of the cesspool cleaning cart simply dumped its contents on the way to the assigned site and because the waste disposal sites did not prevent the refuse from draining into the river—which then was confirmed by a municipal investigation.[16] In their correspondence with the chief of the city police, the Moscow municipal board members wrote: "The different wastes that are constantly discharged into the Yauza from the houses and factories located along its stream have long ago turned the water in this river into a dirty fetid liquid unsuitable for any use. Despite this, no serious measures have been taken against such contamination of the river. . . . This threatens to soon turn this river into the source of infection not only for the residents of the nearby areas, but for the entire city of Moscow, where the Yauza brings its wastes."[17]

The Sewage System

The growing population not only produced more waste but also required more water, and the inadequate system of waste disposal was blamed for the deficit of freshwater in the city—because the water in the wells and rivers became too polluted and unsuitable for drinking. Discussions on the sewage system and on water supply thus developed along the same lines. On the one hand, constructing the sewers was seen as a way to improve the quality of the natural water sources in the city. On the other hand, should the new water supply network be developed, the water usage would grow, so a sewerage system would be necessary to carry away the increased volume of wastewaters, which in turn would facilitate the movement of the sewer cargo in the pipe.

What had changed by the last decades of the nineteenth century, however, was not only the waterways themselves but how Moscow society perceived their pollution. One aspect of this was scientific—connected to the development and institutionalization of medicine, chemistry, hygiene, hydrology, and water engineering in Russia and the spread of scientific ideas about sanitation and disease among the public. The rivers of Moscow became an object of scholarly research, and scientific demonstrations were used to confirm the lay conclusions that they were serving as "the bottomless cesspool" and "cloaca" for the city dwellers. Physicians, chemists, hygienists, and engineers raised awareness of the pollution of the Moscow waterways and alerted the Russian public about the dangers of the unwholesome environment, rotting wastes, and water contamination.[18]

In the 1870s, it was the miasmatic theory that framed understanding of pollution and discussion about waste treatment in Moscow. By the 1880s, germ theory challenged the miasmatic explanation, although it was far from established. As Erismann explained in his university lectures on hygiene at the Moscow Imperial University in the early 1880s:

> the question of the sanitary or, rather, the pathogenic meaning of water, of the role that, in the opinion of many, water plays as an etiological moment in the development and spread of disease, is a very difficult and complicated question. . . . It is generally accepted that the water which contains substantial quantities of organic material liable to rotting should be considered suspicious, either because the products of decomposition can directly cause pathogenic processes or because their presence in water suggests that the latter may contain human feces and, in them, the embryos of infectious disease.[19]

In Moscow in the last third of the nineteenth century, the bad smell of rotting wastes signaled a health hazard, and this connection was not challenged by bacteriological discoveries. Numerous accounts produced by physicians, engineers, city officials, and lay public invoked the stench of Moscow's cesspools and waterways to frame the absence of a proper system of waste disposal not just as a nuisance but as a major health concern. The stench was blamed for high mortality rates and the spread of diseases such as typhus and cholera and even smallpox, measles, and scarlet fever.[20]

Even in scientific circles there was little interest in exploring this perceived causality any further and explaining the exact links between waste, pollution, and disease, despite the introduction of bacteriology and laboratory methods. The acceptance of bacteriological discoveries did not mean the disappearance of the miasmatic approach. Research on the Yauza by one of Erismann's students, Andrei Sokolov, reveals that the miasmatic and bacteriological theories were not seen as contradictory and, in fact, could coexist within the same explanatory model. On the basis of his bacteriological and chemical analysis, Sokolov concluded that the Yauza was severely polluted. He connected these results to the relatively high mortality from typhus, typhoid, and relapsing fever (*tify*) in Moscow districts along the Yauza, but he failed to look for the exact link between the two factors, assuming some natural connection between pollution and disease. Not only his reasoning but also his specific language, with its emphasis on smell, is telling: "The Yauza, receiving all possible wastes, infects the banks and the bottom with the rotting elements and the air with the fetid volatile products of putrefaction and consequently, in combination with the other conditions, has an unhealthy effect on the surrounding area. . . . After entering the city, it rapidly changes in its appearance and becomes dirty and gets an unpleasant smell felt from afar; downstream, next to its mouth, it reaches the maximum of its contamination and becomes unbearably stinking, more resembling the refuse than the river water."[21] The fact that Moscow University awarded Sokolov the gold medal of the Department of Medicine indicates that the faculty committee considered such analysis and argumentation convincing or at least very plausible.

The other aspect in the growing sensitivity to pollution in Moscow was political, connected to the growing feeling of Moscow's backwardness and demands for modernization and reforms. Russian medical professionals and through them large parts of the educated public prioritized Western European systems of hygiene; they equated sanitation with "Europeanness" and progress, and backwardness with "Asiatic filth."[22] Therefore, the "cesspools" of Moscow—meaning both the backyard pits and, metaphorically, the polluted and fetid city waterways—were seen not only as causes of disease, dirt, and disorder but also as symbols of Moscow's underdevelopment and backwardness, resonating with an old image of Moscow as an "Asiatic" or at least a "non-European" city.[23] The Moscow sewage system, the first metropolitan sanitation project in the Russian imperial core, was a task much bigger than merely removing human excrement—it would be a

The Sewage System

symbol of Russian modernization and a way of asserting Moscow's parity with Western European cities.

A European Authority

The construction of sewers was not only the largest and the most expensive project of the Moscow municipality but also the one that took the longest to complete. In the words of the city reformers, the sewage system was for years a "mirage." It was discussed for so long that it seemed almost impossible to achieve, but at the same time the presence of such a large-scale and costly project on the agenda of the city slowed down the implementation of other endeavors.[24] These discussions on the sewage system, however down-to-earth their focus might appear at first glance, were more than just a search for the best technical solution: they reflected the insecurities and ambitions of the reformed Moscow municipality that was exploring its role as a new type of Russian public government.

One interesting aspect of Moscow's path to the sewage system is that its turns do not seem to be directly connected to any major public health crisis. Cholera has often been credited for prompting sanitary advancements in European and American cities. With forty-four cholera years between 1823 and 1914, which claimed more than two million human lives, Russia was the most frequently and most violently affected country in Europe. Yet all the discussions and key decisions regarding the sewerage in Moscow happened during the "calm" years between the mid-1870s and the turn of the 1890s, the longest period without cholera. By the time the sixth cholera pandemic arrived in Moscow in 1892, the sewage system project was already under way.[25] Typhoid, another disease of poor sanitation, remained a persistent problem in Moscow throughout the period. However, the reported numbers of typhoid deaths in Moscow were lower than in Saint Petersburg, Paris, or London, and these numbers were in decline in the 1880s.[26]

The daring idea that Moscow could have a sewage system did not come from the Moscow authorities themselves. In 1874, an engineer named Mikhail Popov submitted to the Moscow City Council a project of sewers that he had devised on his own initiative and at his own expense. The project proposed a sewage system that would cover the entire city territory, excluding the sparsely populated areas, and remove the waste and stormwaters together to filtration fields located on the right bank of the Moskva River, near the village of Kolomenskoe. The price of this sewage system was estimated at sixteen million silver rubles. Popov clearly supported the miasmatic theory of disease and believed in the dangers of rotting wastes and foul odors. The main promise of the sewage system, he claimed, would be the decline of mortality from various diseases, general urban sanitation, and the increase of water supply. He believed that the construction of the sewerage system would improve the quality of water in rivers and wells, which could then be used to compensate for the deficit of drinking water in Moscow.[27]

This proposal came at a time when the reformed Moscow municipality was only testing its new role. It was unclear what its powers and responsibilities were and how its work should be organized and financed, let alone how to embark on such a complex and expensive project. The sewage system sounded like an interesting idea, but would it deliver what it promised? Was Popov's plan feasible? Would it work in the Russian climate where nothing like this existed? Was a Russian municipality actually allowed to do something like this? To answer these questions, Popov's proposal was sent to four different expert committees at all levels of Russia's local and imperial bureaucracy.[28]

Moscow authorities and the Ministry of the Interior received the idea positively and proposed an extensive topographical study to prepare a detailed project. Knowledge of the city terrain, soil, waterscape, climate, metabolism, and population was, in fact, so limited and the program of the necessary research so vast that it took eight years to complete the project. The city commissioned the leveling plan, measured streets and quarters, estimated the density of population and the industrial water consumption, collected meteorological data, inspected the riverbeds and studied the depth of soil freezing in winter. The Agricultural Academy (Petrovskaia sel'skokhoziaistvennaia akademiia) created a test irrigation field to try out how sewage farming worked in the Moscow climate.[29]

Although all the city and state committees found Popov's plan acceptable, municipal leaders decided that opinions of Russian experts alone did not suffice. In 1880 they invited a famous German engineer named James Hobrecht, the author of Berlin's combined sewerage system, to review Popov's project. Hobrecht gave credit to Popov's hard work but castigated his project, concluding that "the fundamental assumptions for calculations do not endure any criticism and can lead to wrong results."[30] The city government turned out to be receptive to this and in 1881 commissioned Hobrecht to provide an alternative project of the sewage system for Moscow.[31]

Why were council members so eager to dismiss Popov's project and to commission Hobrecht instead? For many of them, Hobrecht, who was called "an authority with a European fame," symbolized the European "progress" that was supposed to come to Moscow with the sewage system. One of the deputies at a council meeting advocated the need to invite Hobrecht with the following arguments: "Although we have our local authorities, we came to the conclusion that they need to be checked by inviting a European authority who can say whether what we have done is good or bad and who can grade us."[32] The discussion reveals two important assumptions behind Moscow's municipal reforms at the turn of the 1880s. First is the idea of European modernity as "the gold standard" that needs to be followed. Second is the image of Moscow as a pupil and a backward city that requires foreign help and guidance to become a "modern" and a "European" city. The symbolism of the Moscow sewage system as a mark of "Europeanness," in the eyes of many of the council deputies, could be enhanced through inviting a European celebrity to build it.

From the very beginning, Hobrecht worked in much better conditions than Popov. He could use the city leveling plan and the results of all the topographical studies that were unavailable to Popov in the early 1870s. His information on the population growth and density in the rapidly developing city was more accurate. More important, unlike Popov, Hobrecht knew he would be paid—according to his contract, he received thirty thousand marks. In his project Hobrecht, similar to Popov, offered the combined removal of human wastes and stormwater to filtration fields, which he located on the left bank of the Moskva River, southeast of the city. The cost of the project was estimated at twenty-three million rubles—almost half as much as that of Popov. The most controversial aspect of Hobrecht's project was that he proposed locating the system in areas that were inundated during the spring floods. In his design, the pipelines were supposed to go above the then-existing street level, which would require significant and very expensive ground elevation works, the cost of which was not included in the project.[33]

Both Popov's and Hobrecht's projects included wastewater treatment, which was still not a universally acknowledged necessity nor a norm for the cities that had sewers. The main task of many wastewater systems built in the mid-nineteenth century was to transport sewage outside the city, where it was often emptied into a river or a sea. It was not until the bacteriological era that the need for filtration and treatment was widely recognized.[34] In Moscow, however, sewage treatment projects predated bacteriological thinking.

In December 1882, the two projects were sent for evaluation to the Russian Technical Society, an expert association that included famous engineers and hygienists who found serious faults in both projects but proposed taking Popov's plan as the basis for the new sewage system.[35] Hobrecht's project was abandoned. The money and time spent on inviting a "European authority" did not pay off. Popov, who still did not receive a kopeck from the municipality for his work, was probably celebrating victory as everything pointed in the direction that after all those years his project would eventually go through. Inspired by the defeat of Hobrecht and in hope of finally getting some financial benefits from his work, Popov made an unexpected step. In 1885 he asked the city council to commission him not only with designing the final project but also with implementing it on the principle of a concession agreement. This turned out to be Popov's strategic mistake. He could not predict that the new mayor, Nikolai Alekseev, elected that year would take a radical course toward the municipalization of public services in Moscow, and a private concession for the biggest and the most expensive infrastructural project did not fit into that vision.

American Design

While municipal delegates were debating how to go about Popov's project, a new development occurred in approaches to sewer design. In 1886, at a meeting of the Moscow Society for the Diffusion of Scientific Knowledge (Obshchestvo rasprostraneniia tekhnicheskikh znanii), a municipal engineer named Vsevolod Kastalskii presented a new type of sewer. Where Popov and Hobrecht had designed the combined single-pipe systems then functioning in many European cities, Kastalskii proposed a separate system that dealt with anthropogenic wastes only and let the precipitation run directly into the surface waters.[36]

Today the separate system is believed to have some sanitary advantage over the combined one because of the frequency of combined sewer overflows during periods of heavy rain or snowmelt, when the system becomes overloaded and releases untreated waste directly into surface waters.[37] In the nineteenth century, the separate system was an innovative project, though not exactly Kastalskii's own. He received the idea from the city of Memphis, Tennessee, where the American engineer George Waring Jr. constructed a separate sewage system in 1881.[38]

The idea of separate sewers had originated in Europe in the 1840s. The famous British sanitarian Edwin Chadwick advocated them in the belief that human wastes could be used as agricultural fertilizer and that such use would make the sewage system self-financing, an assumption that was soon proved wrong. The city of London made a choice in favor of a combined system, setting an influential precedent for the design of sewers in many European and American cities.[39] It was not until the turn of the 1880s that the designs of George Waring brought the separate system back to the table. In 1875–1876, he constructed a separate system in the small Massachusetts resort town of Lenox, but it was his work in Memphis that made both him and his design famous. The construction of the Memphis separate sewer system (the design chosen because of its low costs) and the considerable sanitary improvements it caused in a nearly bankrupt city stimulated a great debate in contemporary engineering circles and press. In his influential 1881 report on combined and separate sewers, the American engineer Rudolf Hering concluded that neither system had a sanitary advantage. He recommended a combined system for large densely populated cities and a separate system for smaller cities primarily concerned with household wastes. Until the early twentieth century, American practice followed these guidelines.[40]

The separate system caught the attention of Vsevolod Kastalskii in Moscow probably during the debates of the early 1880s. Moscow, however, was no Memphis. The dramatic difference in size, climate, and geographical conditions of the two cities meant that Waring's design had to be transformed beyond recognition in order to fit Moscow's needs. In the early 1880s, Memphis was a city with approximately 40,000 inhabitants and covered ten square kilometers. Moscow had a population of 750,000 and covered an area of more than seventy-two square kilometers.[41] Memphis was located on the banks of one the world's largest rivers, and Waring had envisioned sewage running into the Mississippi without any treatment. Moscow, on the contrary, had only the small, shallow Moskva River, which could not possibly drain all the wastes produced by Moscow's rapidly growing population. Therefore, all Moscow sewerage projects had to include complex waste treatment facilities.

Both Waring and Kastalskii praised the separate system for its low construction and maintenance costs. Although the combined system, which handled household wastes and stormwater in a single pipe, functioned in many European cities, there were significant differences in climate and precipitation between Western Europe and the United States, where the rainfall was more torrential, or Central Russia, where long-lasting snow cover followed by rapid snowmelt meant that the volume of runoff was highly uneven and seasonal. This had implications for the combined sewage design and operation, demanding larger drains and pipes and carrying a greater risk of sewer overflows. The separate system projects proposed by Waring and Kastalskii excluded the provisions for dealing with volatile levels of precipitation and only needed to discharge human waste—the volume of which remained relatively stable in all seasons. This allowed them to reduce both the size of the pipes and the costs of the project.[42]

However, the main justification for building a sewerage system was the improvement of urban health, and in this matter Memphis and Moscow relied on very different medical-sanitary views. Waring was a major advocate of the miasmatic theory of disease and strongly believed in the idea that "sewer gas" produced by decomposing wastes was the primary source of infectious disease. The sanitary rationale for his system was based on tightly fitting pipes that would protect the households from "poisonous air."[43]

In Moscow, on the contrary, the final argument in favor of the separate system would be bacteriological, although this was not obvious at the beginning. Kastalskii himself praised the system not for the sanitary but for the economic benefits. In his view, the separate system would mean "lowering the sanitary requirements," because the precipitation would not be sent to filtration fields, but it was better to have some sort of sewage system and some form of waste treatment than none at all.[44]

When Kastalskii presented his project to the Society for the Diffusion of Scientific Knowledge, his talk had such a resonance that a special committee of engineers and physicians was created to consider the validity of that design and the sanitary effects it might have. The main point of controversy was the question of what was more dangerous—the precipitation waters going into the river without any treatment or the combined system overflows letting out human sludge. Most members of the committee supported the separate system, but the project had one very influential opponent—Friedrich Erismann, who stood for the combined system, despite possible overflows.[45]

The final resolution of the committee revealed that there was no scientific consensus on the matter. Erismann managed to tailor the main statement to his views: it said that the separate system would allow for the "severe pollution of the river water" and could be permitted in Moscow "only in that extreme case if the prompt organization of the combined system is for some reason impossible."[46] Other members of the committee were dissatisfied with such formulations and submitted additional notes with "separate opinions," criticizing the combined system for leaving a sediment of solid wastes in the pipes in dry weather and thus contaminating city air, for the incomplete removal of stormwater, the risks of inundation, and the high costs of construction and maintenance. Furthermore, as all Moscow sewerage projects included waste treatment facilities, it was argued that the combined system required much larger filtration fields and offered fewer possibilities for sewage farming.[47] The most surprising was, perhaps, the separate opinion of Mikhail Popov who, quite unexpectedly, supported the separate system for its lower costs and feasibility, admitting that its sanitary inferiority to the combined system was rather doubtful due to the overflows of the latter.[48]

Although all the participants in the debate agreed that filth and pollution were dangerous, nobody explained precisely how dumping wastes in waterways could affect human health. This was what made the bacteriological arguments, when they finally came up in the debate, appear so persuasive. New discoveries in bacteriology connected infectious disease to specific microbes and showed that water sources—contaminated with the feces of the sick (for example, as a result of combined sewer overflows)—could play a key role in spreading epidemics such as cholera and typhoid.

When Moscow City Council members discussed the competing sewer designs in 1887, the outcome of the debate was decided by the statement of architect Vladimir Shervud, who called the combined system "the perfect laboratory for bacteria" and blamed it for the spread of infectious disease:

> There is one serious hygienic question that should be considered—that the excrement of people sick with typhoid, cholera and so on do not get into the river. That would inevitably be a disaster. . . . I compared those systems several times, but I have to repeat that the main question is in the dilution of wastes. Does the dilution really eliminate the bacteria? This is the question. Pasteur, for example, proves that it does not. At the recent Hygiene Congress in Austria, Dr. Brouardel argued that infection is transmitted not so much through the air, not so much from the laundry of the sick, but through the water polluted with excrement of those suffering from typhoid and other contagious diseases.[49]

This bacteriological argument highlighting the danger of the combined sewer overflows helped Moscow authorities decide in favor of the separate system, and this was the sewerage type that eventually was built. It would be wrong to assume that the separate system was chosen because it relied on "correct" scientific ideas, however. Historians have reminded us that science turns into policy only when scientific conclusions correspond to the more general political and socioeconomic goals of the decision-makers.[50] Shervud's argument could have been lost if it had not fitted so well with the political agenda of the Moscow mayor Nikolai Alekseev. Although the municipality in the times of Alekseev presented itself as a follower of experts, the absence of a scientific consensus allowed the mayor to turn the scientific discussion to his own advantage and choose the project that best suited his political goals. Alekseev not only wanted to finally move forward with the sewerage project. He had always stood for municipalization; he intended to expand the sphere of municipal authority and bring more enterprises and tasks under its control. The concession proposal of Popov clearly did not fit this plan, as Alekseev argued in front of municipal deputies:

> The city council members many times spoke about the harm of concession; in all the previous cases one could contest it and have a different opinion but it is hardly possible to think that the sewage system can be a subject of concession.

Is there at least one example of the sewage system being the subject of concession, of it bringing income and still satisfying the demands of the city? The sewage system is such an enterprise with which it is unthinkable to expect profit; it brings losses to the city budget but at the same time brings advantage to the house-owners in the financial aspect and to the city residents in the sanitary aspect. . . . This enterprise should be created with city money and should absorb the profit brought by other municipal enterprises, for example, by the water pipe.[51]

The recognition of the hygienic advantage of the separate system allowed the municipality to finally and irreversibly reject Popov's concession plan, which had been until then the most likely sewerage choice. Using Shervud's arguments about the dangers of the combined system, Alekseev pressed the city council to reject Popov's proposal and insisted that municipal engineers be tasked with drafting a technical project for a separate sewage system—the decision that was unanimously approved by the deputies at the same meeting. Furthermore, Alekseev explicitly insisted that the project implementation should be given to the municipal engineers. Plausibly, with commissioning only municipal engineers, Alekseev wanted to retain full control of construction and to avoid the appearance of another "Popov" who would want to push for a concession agreement.[52]

The germ theory thus became an important trump card of the opponents of Popov's project in the course of municipal discussions, in addition to the lower costs and the higher feasibility of the separate system. It was a convincing, handy, and timely argument that allowed Alekseev to reach his goals—to reject Popov's concession project, to continue with municipalization, and to choose the cheapest project while appearing to support the newest scientific ideas. This, however, did not necessarily mean exclusive support for bacteriology; in fact, broader sanitary explanations would continue to be invoked in the later work on the sewage project.[53]

The model of European metropolitan modernity was essential for articulating the need for a sewage system in Moscow, but the ultimate design did not come from Western Europe. The aspiration to catch up with European metropolises paradoxically led Moscow authorities to adopt a design from an American city more than twenty times smaller in size. Not only did that initial design have to be rescaled and adjusted to Moscow's conditions, in order to include a waste treatment plant that was absent in Waring's project, but the entire scientific rationale behind that separate system shifted entirely—from anti-contagionism to contagionism, from miasma to bacteria, from the sewer gas to the untreated waste released into the waterways. In this respect, the Moscow sewage system, designed in the era of major transformations in medicine and sanitary engineering, was part of the global process of validation and establishment of the new bacteriological theory of disease—controversial and politicized as it was.

This story highlights the often overlooked technological connections between the United States and Eastern Europe before the twentieth century and also helps us reassess the role of smaller and peripheral cities as donors of infrastructural innovation.[54] The imperial engineers sometimes referred to the separate system as "the American system" in order to emphasize its difference from the "European" single-pipe systems.[55] The construction of the Moscow sewer system was not just a local project. It was a kind of international experiment, a test for a new technological system—in different climatic, scientific, and social realities—and a step in its further diffusion. As Moscow became the capital city in 1918 and as Moscow's sewerage was and still is the largest urban sewer system in Russia, it had, not surprisingly, a profound impact on future designs. Early Soviet hygienists and engineers called the Moscow system "the main practical school and the model for other cities building sewerage," "the testing station for the study of urban sanitation and water protection" in the country. When many new sewerage projects were initiated in the 1930s, during the Soviet industrialization, it was the separate system that would be the standard design, although that design was no longer referred to as American.[56]

CHAPTER 4

What Happened to Waste?

The crucial decision in favor of the separate sewage system was made in 1887, but another decade had to pass before it was put in operation in 1898. Although this year was the beginning of the sewage system, this is when most historical accounts of imperial Moscow's infrastructure usually stop. After more than twenty years of discussion and construction, the city finally got the sewage system; the rest is of interest only to sanitary engineers and not to historians. However, what did the operation of the sewerage actually mean? Who used it and how? What wastes went in? What happened to them? Did the system deliver what it promised?

Looking at the history of urban infrastructure through the prism of environmental justice—now a classical perspective in environmental humanities—reveals how socially uneven the distribution of environmental burdens and the access to sanitation actually was. This perspective is still new to imperial Russian history. The focus of a lot of existing anglophone scholarship on historical environmental justice in cities is on the discrimination faced by ethnic and racial minorities. Imperial Moscow was racially and ethnically a very homogeneous city—in 1897, 95.0 percent of its population was Russian, followed by 1.7 percent German, and 0.9 percent Polish.[1] But Moscow too was an arena of complex social and environmental inequalities, uneven access and burdens, as well as conflicts over land use that the new wastewater system was embedded in and helped to shape.

Whose Waste?

Let us start with the question of access to sewerage in Moscow. The early sanitation projects of both Hobrecht and Popov planned sewers for the entire city. However, when municipal leaders decided in favor of the separate system, it was Alekseev's idea to have the project built at first only for the central part. Saying that "it is hardly possible to start the implementation of such a grandiose enterprise for the entire city at once," Alekseev wanted to make the project cheaper and more feasible with that decision.[2]

The sewerage construction was divided into several stages, which meant that inequality in access to sanitary amenities was an inherent part of the project from the beginning and for the entire imperial period. The first part—or, as contemporaries referred to it, the first line of sewers, built in the 1890s—was designed for the central area of about seventeen square kilometers, or one-fifth of the city territory, with ca. 400,000 population in 1902.[3] The construction of the second line began soon after and was planned for completion in 1919, but as the works were interrupted by both war and revolution, it was not until the mid-1920s that it was put into full operation. By that time, the city had grown beyond the border the imperial planners had in mind, and the third line of sewers was necessary to serve the new neighborhoods. The first line of sewers covered the area within the Garden Ring, the outer chain of boulevards that followed the line of the former city walls, with several protrusions. These were in the northwest along Tverskaia Street, the main road to Saint Petersburg; in the west along the bank of the Moskva River; in the south along Bolshaia Kaluzhskaia Street, where several municipal hospitals were located; and in the northeast in the densely populated parts of the Meshchanskaia and Basmannaia districts. In terms of physical environment, these central districts were the most built-up part of Moscow, which was also the part that most conformed to the image of a European city, with its rows of multistory stone or brick buildings and paved streets. Throughout the imperial period, the primary construction material in Moscow was wood. Cold Russian winters and short summers favored the use of wood because wooden houses were quicker and cheaper to build, easier to heat, and they did not have the problem of excessive humidity, whereas for stone and brick buildings dampness remained the most common sanitary concern.[4] Yet the wooden houses of Moscow were often looked down on by municipal reformers and regarded as a sign of provincialism, backwardness, and lack of urbanism. In 1890, only one-third of the city's buildings were made of stone or brick, but in some central districts their proportion was more than 90 percent. The dominance of wood as a construction material in Moscow also meant low-rise buildings: just 4 percent of houses in 1890 had more than two floors, and most of those higher buildings were again located in the central districts that would be covered by sewers.[5] The new sewage system promised to make the "modern" city center even "more modern." Not only would it create more private and hygienic homes, appropriate to the growing bourgeois sensibilities, but by solving the problem of waste disposal it would trigger further densification and renovation of the housing stock and the construction of higher and bigger buildings.[6]

Who lived in these districts with access to wastewater infrastructure? The overall social topography of Moscow was very centralized; the historically developed radial-concentric structure of the city endowed the center with the highest economic and social potential. Both the monetary and the symbolic value of Moscow districts decreased gradually from the center to the city border. Housing prices could serve as an indicator of this pattern: in 1890, the average annual price per room was 95–161 rubles in the northern downtown districts within the Garden Ring and 87 rubles on the southern bank of the river (central Zamoskvorechie). The range was 67–98 rubles in the areas immediately outside the Garden Ring and 43–68 rubles in the districts lying further out.[7]

It would be natural to expect that this economic and symbolic value of the central zone meant that the poor were driven out toward the city border and that the downtown was appropriated by the upper and middle classes. In fact, although the outskirts were indeed populated by the poor, it appears that Moscow's working population was spread out all over the city territory. No district of Moscow was restricted for or even principally inhabited or used by the middle class. Even in the quarters with the highest land and apartment prices, around Tverskaia Street and the Kremlin itself, the working class accounted for almost one-half of the population. One of the important factors of Moscow's social topography was the tight connection between place of work and residence. Low intra-urban mobility and an undeveloped public transportation system created the necessity of living close to the workplace. As a result, Moscow did not have horizontal social segregation, which at the turn of the twentieth century was becoming visible in the spatial organization of Western metropolises. Horizontal differentiation existed, yet it was expressed not in the form of class segregation but, rather, in the presence or absence of the upper classes.[8]

This meant that the sewer system covered the areas where the wealthy tended to live, but it was not intended to serve the wealthy alone. Not only was it supposed to provide sanitary amenities to the large numbers of commercial and industrial workers living in the central districts, it also covered the most notorious city slum around the Khitrov market and another

TABLE 4.1. **Number of properties connected to the Moscow sewage system by year**

1898	219	1905	296
1899	849	1906	191
1900	902	1907	203
1901	501	1908	202
1902	451	1909	242
1903	431	1910	263
1904	329	1911	368

Source: Nikitin, "Kanalizatsiia goroda Moskvy."

ill-famed neighborhood called Drachevka. One can say that, in its original plan, the sewerage was a rather inclusive project aimed at bringing environmental sanitation to a diverse urban population.

However, if we zoom in from the district level to the individual properties, the picture is more complicated. The realization of the environmental and public health goals of the Moscow municipality depended not only on where the sewerage was built but also on whether and how it was used. Overall, the project was meant to serve 6,785 properties of the central districts. By the end of 1898, the year when the first line of the sewerage was launched, only 219 properties were connected to it. Ten years later, two-thirds of intended properties were using municipal sewers, and after the peak years of 1899–1900, the speed of connection declined (see table 4.1).

There were several factors explaining these dynamics. The first was the relative weakness of Russian municipalities. The connection to sewerage in Moscow was not mandatory, and the municipality had no legal power to force property owners to join the system. Although it petitioned the central government for the right to introduce the compulsory use of sewerage, the request was declined. It was not until 1912 that municipalities in Russia were empowered to make connection to their sewer systems obligatory.[9]

The second problem was that this new sanitary amenity was expensive. Owners had to pay 3 percent of the property's net revenues (*dokhodnost'*) for connection to the sewerage and 4 percent annually for its usage. The upper limit of these fees was fixed by the Ministry of the Interior, which again signals the limitations of control that Russian municipalities had over their own enterprises. However, in reality, these fees were not sufficient to cover the maintenance of the sewerage, especially when only a small part of intended properties were paying them; for the first eight years the enterprise was operating at a loss. Although, back in 1887, Alekseev believed that the

Table 4.2. Financial Year Results of the Moscow Sewage System (in rubles)

1899	−371,000	**1905**	−162,000
1900	−329,000	**1906**	−267,000
1901	−71,000	**1907**	158,000
1902	−112,000	**1908**	187,000
1903	−125,000	**1909**	293,000
1904	−211,000	**1910**	377,000

Source: Nikitin, "Kanalizatsiia goroda Moskvy."

removal of wastewater could not be a profitable business and its goals were public health and not financial gains, in the early twentieth century, city authorities were not ready to put up with such a state of affairs.[10] In 1906, the municipality petitioned the imperial government to increase the maximum fee for the usage of sewers to 5 percent, which eventually made the enterprise profitable (see table 4.2).

Finally, another factor that slowed down the connection of private houses to the sewerage was the fear of a system overload due to industrial refuse. Although, in the 1880s, about one-fifth of Moscow's factories were located within the Garden Ring, the sewerage commission concluded that the decontamination of industrial wastewater was not a municipal concern. In the early 1900s, the municipality found out that in fact it was, and that the sewerage had to receive factory wastes as well. In 1903, it became evident that the capacity of the system was reaching its limits, despite the fact that only a small part of planned properties had been connected. In this situation some noncentral estates were denied access to use the sewers until the expansion of the waste treatment facilities.[11]

The nonobligatory connection and its high costs meant that usage of sewers within the covered districts was uneven. In 1905, only about one-half of downtown properties were connected to the sewage system, but this one-half produced 80 percent of all property revenues and housed three-quarters (425,000) of its population.[12] The owners of bigger, renovated, and more expensive properties were more likely to invest in connection to the sewerage, while the availability of indoor plumbing attracted new and wealthier tenants and led to the further increase in rent. Many of the smaller, older, or lower-quality houses where the downtown poor were likely to live continued to rely on outdoor cesspools, privy vaults, and sewage

FIGURE 41. Water closets at the municipal night shelter. From Uspenskii, *Moskva*. National Electronic Library (Russia).

carts until the eve of the First World War, as did the residents of the outer city districts.

Connection to the sewerage transformed sanitary facilities inside the properties, but again there were differences among them. Municipal rules required all connected houses to install water closets and kitchen sinks. Privies, cesspools, and similar arrangements were forbidden, so, once the property was connected to the sewerage, all its residents, wealthy or poor, masters or servants, were to use the water closet. The necessity of installing a water closet required abundant water supply, either from the water pipe or from streams or wells. According to municipal regulations, all new or rebuilt houses had to organize a separate, warm toilet in every apartment—a significant rise not only in standards of sanitation but also in standards of privacy compared to the facilities most Muscovites had been using before. However, older houses where the organization of private toilets was not immediately possible were allowed to have one water closet for several apartments or a separate toilet in the backyard to be used by all tenants. Shared toilets were also a norm in municipal housing facilities for the poor (see figure 4.1). Meant to reduce the financial burden on property owners and encourage the usage of the sewerage, this policy resulted in uneven sanitary standards across connected properties. In 1912, Moscow municipal engineer Iakov Zviaginskii described the shared backyard toilets:

> They are usually organized in a least appropriate spot, in a nook of some kind, without any light or ventilation, and the results are quickly felt. The facilities turn bad in a short time. The devices are not washed, and excrements accumulate on the floor, producing intolerable stench and serving as hotbeds of infection. This condition of backyard water closets is helped by the fact that the eye of the owner or manager rarely turns to the water closet, and they are left up to janitors, although the proper upkeep of backyard water closets should be vigilantly looked after.[13]

The shared toilet was intended as a temporary solution that would disappear with the gradual renovation of the housing stock, but it remained an unavoidable sanitary reality for many poorer residents of the central districts until the end of imperial period and for years thereafter.

In 1911, 5,400 properties with 660,000 population had access to the sewerage, sending twenty-one million cubic meters of wastewater to the treatment facilities that year. By 1925, more than 8,000 housing units with 930,000 population were connected to municipal sewers, most of which had been designed and built before 1917. The numbers of people who benefited from this public health project of the Moscow municipality are impressive, but even more impressive perhaps are the numbers of those who

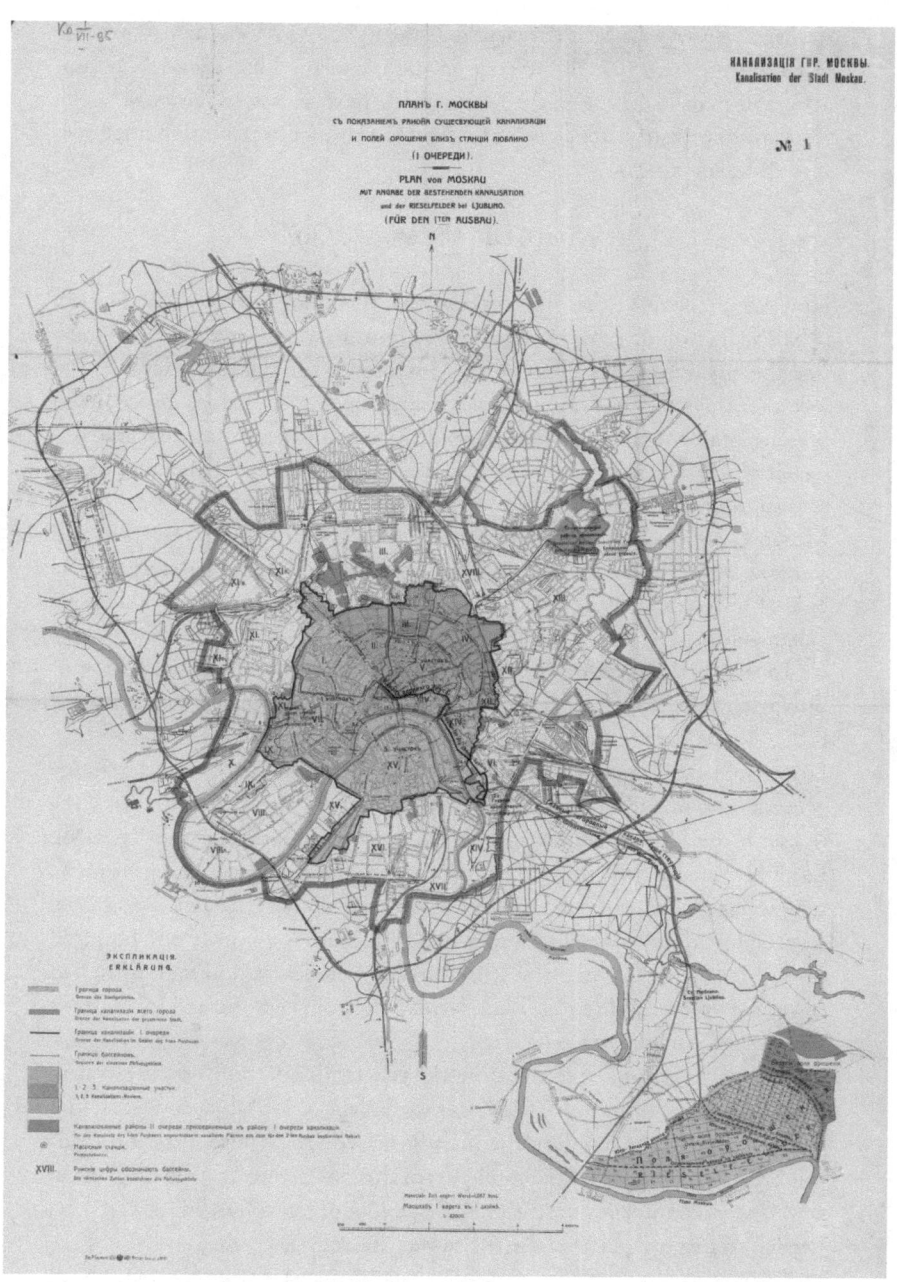

FIGURE 4.2. Plan of Moscow showing the area covered by the first line of the sewage system and the filtration fields in Liublino. From *Kanalizatsiia goroda Moksvy [Karty]*. National Electronic Library (Russia).

were left out, especially as the city continued to grow. More than 70 percent of Muscovites in 1902, 60 percent on the eve of the First World War, and about 50 percent in the mid-1920s did not have access to sewerage.[14] The Moscow of private water closets and the Moscow of cesspools remained two very different worlds.

Where Did the Waste Go?

Now let us move to the other end of the pipe and look at what the new urban infrastructure meant for the communities and environments outside the city, where the cargo of the Moscow sewage system ended up. The idea that the sewer liquid should necessarily be cleaned before disposal was present in all Moscow's sewerage projects from the beginning and predates bacteriological thinking. The explanation lies probably in the absence of any appropriate "sink." Locked in the middle of the East European Plain, Moscow was too remote from the sea or any other large water reservoir, and the slow Moskva River with its weak flow and populated riverbanks, in the eyes of sewerage planners, could not provide an adequate outlet to deposit urban waste.

In the final project of the Moscow sewage system, the chosen option of waste neutralization was intermittent soil filtration combined with a sewage farm. At the end of the nineteenth century, sewage irrigation and sewage farms were gaining popularity in Europe, as a way of processing the sludge accumulated in urban sewer networks, and were also receiving some interest in the United States. In contrast to sewage farming (that is, using human waste as agricultural fertilizer, which has a centuries-long history), new sewage farms came to represent a specialist activity, designed primarily to process urban waste. In Britain in 1880, there were some one hundred sewage farms. In Germany by 1910, forty-nine cities, including Berlin, had sewer networks leading to sewage farms, most of them located in the eastern part of Prussia.[15]

The sewer liquid, collected from the central districts of Moscow, was to be transported to filtration fields near the village of Liublino, southeast and downstream of the city (see the map in figure 4.2). Those filtration fields required a lot of space. If the total area of the city covered by the first line of sewers was seventeen square kilometers, a plot of ten square kilometers was needed to dispose of the waste the sewers collected.

This was, however, not a no-man's-land. Most of the land intended for the filtration fields (eight and a half square kilometers) belonged to the peasant communities of the Moscow province, with the rest owned by the imperial court and local churches. Although this land might have appeared empty to municipal experts who identified it as the destination for the sewer cargo, it played a vital role in the subsistence of the peasant communities and their entire way of life. At the turn of the twentieth century, Central Russia, including the Moscow province, suffered from a lack of land because of the unfavorable conditions of peasant emancipation, high birth rates, and the persistence of the peasant commune as the main owner of land. This meant that individual peasant allotments, redistributed regularly by the commune according to the number of family members, were barely enough to subsist the population living from the land to subsist on.[16]

According to the evaluations of the Moscow zemstvo, which was not even consulted by the government commission supervising the construction of sewerage, the filtration fields of the sewage system required expropriation "for public needs" of more than 11 percent of the land belonging to the affected communities.[17] Of the twenty-four villages, two would lose more than 40 percent of their land and another three would lose more than 20 percent. Given the already severe scarcity of land, the loss of such a considerable area, the zemstvo argued in 1896, threatened the existence of the entire peasant communities because the remaining land would not be sufficient to subsist the village population:

> The meaning of this matter is that the land given to peasants "for their subsistence and the performance of their responsibilities in front of the Government" is the main condition of their existence. That is why depriving them of land or a part of it, without which it is no longer able to "subsist them," cannot be reimbursed by any monetary compensation. . . . The damage induced by the destruction of the basis of your subsistence, by the impossibility to continue living from agriculture, that forces you to involuntarily change the way of life or a place of residence cannot be reimbursed because there is not and cannot be a scale to evaluate the material loss and because moral interests are also affected here.[18]

Even in the less affected villages, the expropriation would undermine existing agricultural practices. The filtration fields would stretch along the bank of the Moskva River, cutting the rural population off from the major water artery in the area. Some communities would lose seemingly a small portion of land, but it was the only plot that could be used for pastures and for making hay to feed the livestock during the long Russian winters. Given the population density in that part of the Moscow province, it was impossible for peasants to find alternative plots for these purposes. Furthermore, the land owned by the imperial court had been rented by peasants and used to graze cattle and make hay, but this too would be taken away.

The lack of pastures and, consequently, the lack of livestock in the countryside of Central Russia had been a long-standing concern of the Moscow zemstvo. Zemstvo experts regarded it as a cause of bad harvests because of the lack of natural fertilizers and a cause of rural poverty that undermined zemstvo's own economic base.[19] It is not surprising that the prospect of losing already scarce meadows in the province led to the zemstvo's mobilization against the project of filtration fields.

Outlining the detrimental consequences of the planned site for Moscow's wastes for the rural communities, the zemstvo proposed either moving the sewage farm to a less populated area or preserving the land under the filtration fields in the peasant property and letting it be used as pastures and meadows, which the local villages desperately needed. This plan tried to reconcile the interests of both sides and to retain peasants' agricultural practices. The zemstvo also requested that its representatives and a zemstvo sanitary physician be included as full members in the commission supervising the construction of sewerage. Zemstvo officials argued for the creation of a separate body that would mediate the conflict between the peasants and the city. Such a compromise, however, did not fit the vision of a modern scientific enterprise that municipal experts were determined to establish.

The person commissioned to answer to the zemstvo was a professor at the Moscow Agricultural Institute, Vasilii Williams, the organizer and future director of the filtration fields. The son of an American engineer, Williams would later become an iconic figure in Soviet soil science, an active supporter of Trofim Lysenko, and one of the founders of Michurinist agrobiology.[20] In his response to the zemstvo, Williams claimed that the filtration fields and the sewage farm were too complicated enterprises to involve peasants in any way. In his view, the practices of grazing and haymaking that existed in the peasant communities were so outdated that they would only ruin the entire system of waste treatment, turning the filtration fields into "rotting swamps that would pollute both the local environment and the main river of the province."[21] Williams denied that peasants had the ability to change and adjust their mowing patterns to the operation of the sewage farm. At the same time, he seemed to ignore how much adaptation the transformations brought about by forced alienation of land would require from the local communities. The argument that peasants would lose all their pastureland was not convincing to Williams. He advised peasants to "make an easy step towards intensification and ideal suburban economy" and completely abandon grazing cattle, keeping it stabled all the time instead. In stable, Williams claimed, "without moving, it would receive the luxurious grass" instead of "wasting power and energy on walking through the meadows in search of food."[22]

FIGURE 4.3. Filtration fields in Liublino. From *Al'bom zdanii*. National Electronic Library (Russia).

The alleged "backwardness" of peasants is a familiar trope in Russian history. In late imperial Russia, this assumption was held by a wide spectrum of social elites who believed that "backward" peasantry needed the intervention of those who saw themselves as "progressive."[23] This vision was shared by the zemstvo activists, too, as they assumed the right to speak about and for the peasants and defend their interests in the conflict with the city. Whereas the Moscow municipality claimed to promote the public good of the urban population, the zemstvo of the Moscow province had a similarly ambitious mission in relation to peasantry, and here two quests for public good clashed. In the words of the zemstvo representatives, "in this case the serious public needs of Moscow crush the interests that are not private but also public, and of equally serious character because it is impossible not to recognize the destruction of the base of subsistence and economy of entire settlements as a matter of public interest in the strictest and most serious meaning of the word."[24]

The strategic initiative in this conflict was on the side of the city. For Williams as well as for municipal and state experts, the public good of urbanites was far more important than the interests of peasants (not to mention animals). The rural environments and the communities living from them had to yield and transform in order to meet the environmental demands of the growing modern city. Furthermore, this process was presented as a boon that would bring modernity and reason to inert peasants who, in the eyes of their urban observers, were otherwise unable to organize a functioning agriculture.

The arguments of Williams determined the decision of the government commission supervising the construction of the sewage system. Not only did it decide to support the municipality, it even refused to admit zemstvo representatives to its hearings. The commission's resolution concluded that "the sewage farm, correctly organized on the filtration fields by the Moscow municipality, will not only improve the economic condition of peasants but will also be a vehicle of the proper culture among peasants of the Moscow province."[25] It would be not just human excrements that the sewage system would carry to the countryside but with them urban science and progress to the "backward" peasant world.

In April 1897, Emperor Alexander III signed a decree on the alienation of lands for the needs of the Moscow sewage system, and the next year the filtration fields received their first portion of urban waste (see figure 4.3). Initially, municipal experts hoped that the agricultural output of the sewage farm might even generate some profits, but it soon became clear that those expectations were naïve. In his report as director of the sewage farm, Williams explained the lack of profits by the need for numerous complicated works not related to agriculture per se. The high volume of human waste coming to the filtration fields meant that only a small part of its territory (about seventy-five hectares in 1900) could be used for farming. In addition, most of the land under the filtration fields had previously been used as pasture, and the sewage farm management had a severe problem with weeds that thrived on the soil fertilized by human excrement.[26]

The choice of potential plants was also quite narrow, and their sale prospects dubious. Agricultural production depended not on market demands but on how well the plants could tolerate the sewer liquid. The Moscow sewage farm had to concentrate on production of forage crops and vegetables such as cabbage, beetroot, cucumbers, potatoes, and onions. Sewage farms operating in Western cities often struggled with selling their vegetables because of concerns over food hygiene. Even though many experts claimed that products of the sewage farm were safe, local consumers detested the idea of eating something that had grown on or in proximity to human feces.[27] Moscow found an ingenious solution: vegetables from the sewage farm were sent to municipal hospitals whose working-class patients were unlikely to inquire about the origins of their food and hardly had a choice anyway. The nutrients that left Moscow in the form of excrements, mixed together with potentially toxic industrial waste, returned to the city on the plate served to the urban poor.

In addition to sewage farming, a part of the filtration fields was left under meadows. As Williams described in his report: "the best hay" from those meadows was cut by municipal workers and fed to the cattle at the sewage farm, while about 140 hectares "with bad hay" was mowed by local peasants for one-half or one-third of the harvest.[23] This was what peasant communities received in replacement of the expropriated land.

Despite the words "fields" and "farm" in the name of the enterprise, this was an industrial project in its scale, purpose, and material and social organization. The project relied on a complicated technological infrastructure, with a network of pipes, drains, and pumps as well as two scientific laboratories—one chemical and the other biological-bacteriological. In 1904, in addition to soil filtration, the city started exploring the option of biological waste treatment at the fields. After successful experiments with the small biological station, in 1908 the Moscow municipality opened a larger facility with a capacity to process six thousand cubic meters of sewer liquid daily. In 1910, municipal engineers developed a project for yet another biological station ten times bigger that could potentially process the entire volume of waste coming from Moscow (in 1909–1910, the average daily intake of the filtration fields was about sixty thousand cubic meters of sewer liquid). That year the city also bought a plot measuring fifteen square kilometers near the village of Liubertsy to create new filtration fields for the second line of the sewage system.[29]

Life and work on the filtration fields was organized in line with that of industrial enterprises. Managed by a small number of experts and administrative staff, the project employed more than two hundred people, each of them having a narrowly specialized task in the agricultural or technical department. The agricultural work was performed by hired people who had no relation to the land they were working on. As at most Russian factories, all the personnel lived on-site, which in that case meant right on the filtration fields. At the northwestern edge of the sewage farm, there was a so-called estate with offices, workshops, barnyards, and living quarters for its personnel: a separate house for the director, an apartment building for the white-collar employees, and barracks for the workers—a material structure that signified "modernity" for its proponents but, at the same time, so vividly embodied the hierarchies of the Russian society in spatial forms.[30]

The sewage system was Moscow's most expensive project aimed at public health and public good—and also the largest in scope and achievement. Together with its filtration fields and biological stations, the sewerage gave the city a sanitary advantage for many years ahead. As late as the 1950s,

Moscow was Russia's only large city where over 90 percent of collected urban waste underwent treatment.[31] Even though not in full operation before the Bolshevik revolution, the sewers of the imperial period literally created a base on which the Soviet authorities could build their capital of socialist modernity.

The sewage system brought new standards of hygiene, comfort, and privacy to thousands of Moscow homes, but it also reflected and reinforced the city's geographies of privilege. Although sewerage was meant as an inclusive public good and public health project for the city, more than half the urban population in imperial Moscow could not access the new sanitary amenity. The sewage system not only perpetuated but strengthened inequalities both between the center and the expanding and increasingly populated outskirts and, within the center, between those who had access to modern facilities and those who did not.

Even more disturbing, perhaps, is how readily the public good of urbanites (however uneven) was put above the public good of peasants, their property rights, their way of life, and the environments they lived in. With the largest site for waste disposal by their side, they could not partake in the new sanitary regime brought about by the sewage system. The peasant communities would disappear before the water closet would arrive in that part of the Moscow province. The needs of the growing industrial cities and their demand for resources were almost unquestionably prioritized over the needs and concerns of rural communities, which were deemed backward and inert, and the conflict between the two sides was resolved not in the courtroom but by the intervention of the central state. This pattern would be taken to its tragic extreme in the Soviet period, when the peasant world was crushed to provide resources for industrialization, but the germs of that dramatic change had been there before.

CHAPTER 5

The Conundrum of Industrial Waste

In 1899, the year after the Moscow sewage system was launched, Chief of Moscow Police Dmitrii Trepov considered industrial pollution of waterways a "solved question." The solution, as he described it in his report to the Moscow Governor-General Grand Duke Sergei Romanov, was simple and obvious: the government had to make all the factories connect to the municipal sewers and forbid and close all existing drains into channels and natural waterways.[1] In his post as the chief of Moscow police (1896–1905), Trepov would play an important role in linking municipal reforms in urban sanitation to the broad questions of industrial policy in Central Russia.

How did it happen that the Moscow sewage system, created as an alternative to cesspools for the disposal of domestic waste, came to be seen as the answer to the burning question of industrial pollution in one of Russia's largest factory centers? What was industrial water pollution in the first place and how was it treated by late imperial law, experts, and policymakers? How did industrial pollution relate to the public health paradigm in imperial Russia?

We know how important perceived environmental threats, including industrial discharge, were for triggering urban public health reforms in Moscow and in many other cities. But what about the reverse relation? I argue that advances in the municipal sanitary infrastructure had a profound impact on regional and national policy on factory waste, although connections between the public health movement and concerns with industrial pollution were more complex and contingent than might appear at first sight.

Legal Context

The rapid industrial and urban growth in the two decades after the abolition of serfdom put a major strain on the environment and, especially, on water resources. In the 1880s, accounts discussing industrial pollution of rivers and the dangers posed to human health and well-being became increasingly prevalent Poisonous substances and organic waste in industrial refuse, the imperial experts warned, rendered water unsuitable for consumption and domestic use.[2]

This period also marked the beginning of Russian factory legislation. Four factory laws (1882, 1884, 1885, and 1886) prohibited the employment of children under twelve and night work for women and teenagers, made provisions for the education of child workers, regulated conflicts between industrial workers and their employers, and established a system of factory inspection.[3] Taken together, these laws imposed perceivable restrictions on factory owners, particularly in the matter of labor relations. They also created a climate where further regulation of industry, including the mandatory treatment of waste, seemed like a logical possibility.

Imperial Russia did not have any unified legislation on water pollution comparable, for example, to the Rivers Pollution Prevention Act of 1876 in Britain. This did not mean that the tsarist government made no attempt to protect water resources and that the questions of industrial discharge and water pollution were not legally regulated. Rather, their regulation was dispersed across several legal statutes and decrees and often in unclear, repetitive, or somewhat contradictory formulations, which meant that even contemporary bureaucrats and experts found it difficult to apply.

The basic principles of water protection were stipulated in state legislation such as the Medical and Building Statutes and the Statute of Industry (*Ustav vrachebnyi; Ustav stroitel'nyi; Ustav o promyshlennosti*). The Medical Statute forbade "contaminating water in places where it was taken for internal consumption by throwing into it harmful substances or in any other way" (ruled in 1871) and obliged local police and municipalities to ensure that "rivers and springs in towns and villages were not polluted."[4] The Building Statute and the Statute of Industry prohibited the construction of "mills and factories harmful for the purity of air and water in towns and upstream of towns."[5] This norm was inherited from the early nineteenth century, and its interpretation and application proved difficult in the later contexts of urban and industrial growth. In 1904, the Senate had to clarify that it applied only to particularly dangerous or poisonous industries, while all other factories could be allowed on condition of proper waste decontamination.[6]

State legislation was supplemented with local regulations and decrees, issued by provincial governments, zemstvos, or municipalities. These institutions were responsible for public health and sanitation in their respective regions and controlled the cleanliness of natural waterways and sanitary conditions at industrial enterprises. The permission of a provincial government—or, in some cases, of a municipality—was necessary to open a factory, and this gave local authorities a mechanism to influence factory owners and make them comply with sanitary norms. Local decrees of many zemstvos prohibited littering, contaminating, or dumping and draining waste into natural waterways that were used by the local population for domestic needs or for watering the livestock. These regulations were particularly strict in regard to draining fecal human waste either from residential houses or factories, while formulations about liquid industrial waste were more lenient.

Legal sanctions against violators of water protection norms were formulated in the Statute of Magistrates (Ustav o nakazaniiakh, nalagaemykh mirovymi sud'iami). Prepared as a part of the judicial reform of Alexander II in 1864, this statute distinguished between the crime of illegal discharge of waste and the crime of water contamination. It stipulated a one-hundred-ruble fine for connecting household cesspools or factory drains to city gutters and channels without the necessary permission, and a two-hundred-ruble fine for an illegal discharge of domestic or industrial waste into rivers and streams, regardless of its impact on the waterways. Altering water quality was punished separately. A small fine of ten rubles was imposed for littering waterways with stones, sand, or other substances that "could not cause the contamination of water."[7] Contaminating water used for human consumption or watering of animals was prosecuted according to article §111, which in its original formulation of 1864 prescribed a fine of up to twenty-five rubles or a seven-day prison sentence.[8] However, following the revisions of 1893 and 1906, the maximum punishment was increased to one hundred rubles or a one-month sentence and, in cases where polluted water became harmful for humans, to three hundred rubles or a three-month sentence.[9] Such a considerable toughening of the punishment for water pollution may indicate growing public attention to the problems of industrial waste and river protection toward the turn of the twentieth century.

There are several remarkable aspects of these regulations. The first is their narrow utilitarian approach. Water pollution was seen as a local problem, and waterways were deemed worth protecting only if they were used by the local population for domestic needs or livestock raising. This excluded not only the value of riverine environments per se but also their importance for other social and economic purposes such as fishing, agriculture, recreation, or industry itself, although contemporaries were clearly aware of these connections.[10] A second aspect is the lack of distinction among the violators. The law treated individual house-owners draining their domestic waste and industrial giants discharging thousands of cubic meters of wastewater every day in a similar way. Most important of all, perhaps, was the striking vagueness of the legal formulations: they did not specify what exactly littering (*zasorenie*), pollution (*zagriaznenie*), or contamination (*porcha*) was, which waters were considered dirty, how precisely the dirty waters should be handled, and the degree of decontamination and purification to be achieved in order to discharge treated waste into rivers without breaking the law. In other words, the law obliged factory owners to obtain a set of necessary permissions for draining industrial waste into rivers, but those permissions did not guarantee they would not be persecuted. At the same time, the law required waste decontamination, without explaining what was expected from an industrialist and how it could be accomplished.

This lack of clear definitions and explanations satisfied neither the industrialists nor the advocates of water protection, making both sides debate the regulation of industrial pollution in the last decades of the nineteenth century. In the mid-1890s, the Ministry of Finance, then headed by the influential liberal minister Sergei Witte, attempted to address the issue of industrial pollution. Witte generally opposed any practices that could hinder Russia's industrial growth, and his attention to the questions of waste treatment and its regulation was part of a broader effort to create a coherent legal framework on the organization of industrial enterprises and their supervision. In 1896, the ministry's commission, which also included leading hygienists, economists, and factory inspectors, proposed that the punishment should be inflicted not for the fact of water contamination but for the absence or inadequate use of the waste treatment facilities prescribed by the law.[11] However, this formulation—which would make legislators (and not industrialists) responsible for finding an effective technological solution—was not translated into a legal norm.

The Russian Empire was not alone in its endeavor to resolve the questions of water pollution or to at least specify what it was. At the turn of the twentieth century, other countries were struggling with the same problems; in fact, a developed and specific legislation on water pollution was rather an exception than a norm. In Germany, water laws remained largely the responsibility of the regional authorities, which could choose to "sacrifice" rivers for industrial needs—as happened, for example, in the Ruhr region.[12] In France, altering water quality was criminalized but mostly went unabated because of the lack of adequate measurement facilities, while the adoption of a comprehensive legislation on pollution was repeatedly opposed by the French Parliament.[13] In the United States, the first federal legislation on industrial waste—the Refuse Act of 1899 in the early twentieth century—was used to preserve waterways for navigation rather than to prevent environmental damage.[14] Even in Britain, which had comparatively advanced and coherent legislation on the subject, there were still no clear legal definitions of pollution. The Rivers Pollution Act of 1876 required treatment of "poisonous, noxious, or polluting liquid" before discharging it into a watercourse using "the best practicable and reasonably available means."[15] However, the existing technological solutions were often inadequate, and it was not until 1912 that quality standards based on chemical and biological measurements were proposed for the effluents and river waters, and even then, their implementation was hindered by the onset of the First World War.[16] There was thus no successful foreign model for Russia to follow easily and no international standard for it to adopt. In this situation, it was the search for and the negotiation of definitions of pollution and waste treatment that would shape the practices of waste management in Russia in the last imperial decades.

Moscow Authorities Deal with Industrial Pollution

There were two important milestones in the history of industrial pollution as a political issue on the agenda of Moscow's regional and city authorities. The first one was the cholera epidemic of 1892. The second was the launching of the Moscow municipal sewage system in 1898. Neither of the two was directly related to industry, yet they were crucial to how industrial pollution was framed and tackled.

The 1892 cholera epidemic was a test for Moscow's reformed public health system. In the decade preceding the epidemic, the municipality managed to unite under its umbrella most of the city's health services and to build a reputation as a successful health administrator. The city government also cooperated closely with physicians on every level and strongly relied on medical expertise. In the moment of emergency, this helped Moscow avoid the mistakes that led other Russian cities to disaster: the confusion of the health administration because of the dual structure of power, the failure of local officials to assess the medical situation, their inadequate communication with medical professionals, and the lack of any unified strategy that would allow the disease to be contained.[17] As a result, Moscow arguably passed that test successfully: the epidemic took a moderate toll (fewer than 3,000 cases and 1,450 deaths compared to, for example, almost 5,000 deaths in Saint Petersburg and 7,600 in Hamburg) and did not cause any social unrest, although in some other cities of the empire it provoked violent riots.[18]

The measures employed by the city against the disease stood up to contemporary international standards, developed in the light of the bacteriological discoveries of Robert Koch. The key role in the transmission of cholera was ascribed to contaminated drinking water. It was also known that the spread of the epidemic is in some way helped by human contact. Therefore, the proposed measures concentrated on protecting water sources from human waste, preventing the use of untreated water, stricter control over food products, and the quick identification and isolation of the sick.[19]

Although industrial pollution was not seen as a specific threat in spreading cholera, in 1892 the Moscow municipality launched an inspection of the city's factories to detect and prevent possible sources of water contamination. Inspectors discovered that absolutely all of the checked factories released their waste into the city's waterways and that none of the existing filters fulfilled the task of waste decontamination. In other words, all of Moscow's factories were breaking the laws of water protection. The universality of violations meant that nobody could be punished, and factory owners were only advised to ask municipal engineers for a recommendation of a proper filter. However, when factory owners did inquire, the engineers could not give any definite answer. Neither could the Council of Trade and Manufacture of the Ministry of Finance, which admitted that this question had been discussed all over the world for decades, and no satisfactory solution had yet been found. As a result, there was not much the city could do against the industrial discharge. According to the order of the municipal board, the drains of the Bakhrushin leather factory were closed, but this decision was revoked several months later on the grounds that all other factories were operating in equal violation of the law.[20]

This was how the public health concern and the fear of a disease not related to industrial pollution brought the issue of factory waste onto the table of the authorities. The municipality was not the only political actor involved, however. The cholera epidemic also led to the sudden interest of the local imperial administration in the problems of pollution and industrial wastes in particular, and, more importantly, triggered the application of repressive administrative measures otherwise reserved for political threats in the conflicts over pollution (discussed in chapter 2).

The Conundrum of Industrial Waste

The question of what to do with polluting Moscow factories, raised during the cholera epidemic, remained pending for several years until it caught the attention of Chief of Moscow Police Dmitrii Trepov. In 1893, Trepov brought the question of river pollution up to Grand Duke Sergei, using the pending cases of the Albert Gübner textile factory and the Khamovniki brewery, which were reported to pollute the Moskva River:

> There can of course be no other decision on these reports than to act according to the law, that is to say, to send the police and medical protocols to the district court whose only legal decision can be to close the discharge that in most cases practically equals the closure of the factory. It is known, however, that the absolute majority of factories are in the same conditions as the Gübner factory and the Khamovniki brewery, that is why their legal persecution on exclusively legal basis can have extremely serious consequences: on the one hand, it can lead to the bankruptcy of the factory owners, on the other, it will leave completely innocent workers without their wages.[21]

Leaving thousands of Moscow's workers unemployed was clearly against the interests of the imperial administration and the police, given the growing political mobilization and strike movement of the workers. Furthermore, in 1898, Trepov was considering a plan of gaining workers' loyalty through legal pro-governmental organization to resist revolutionary propaganda and political protest. This plan was developed by the director of the Moscow security police Sergei Zubatov, who believed that the state needed to side with workers and support them in their economic struggle against industrialists.[22] In this situation, combating industrial pollution had to proceed without causing any unrest among workers.

To find an acceptable solution to this sensitive problem, the Commission on Industrial Discharge in the Moskva River (Kommissiia dlia izuskaniia sredstv k preduprezhdeniiu zagriazneniia Moskvy-reki spuskaemymi v nee s fabrik i zavodov otrabotannymi vodami) was formed. The commission brought together Moscow bureaucrats, engineers, and industrialists who were supposed to formulate the guiding principles of how the questions of factory discharge should be handled. The creation of this commission coincided with the launching of the Moscow sewage system in 1898. This new urban technology, as the members of the commission soon assumed, seemed to offer the best practical and legal answer to the ongoing controversy. Commission members thus concluded that to stop polluting rivers factories needed to connect to the sewerage that would process their waste. In practice, this meant that factories in the areas covered by municipal sewers were obliged to connect to the system, while those outside that zone were temporarily allowed to continue draining their waste into the rivers, but any increase of production or any opening of new factories was prohibited until the expansion of the sewage system.[23]

Trepov's own position on the question was even more strict. In his report to the Moscow governor-general, he proposed not only to stop issuing any new permits for industrial discharge into the natural waterways but also to close all existing drains that let out polluting effluents upstream of the city. "Such measures, although serious," the chief of Moscow police maintained, "would not put the industrialists in a deadlock because they would always have the possibility of organizing at their factories, located beyond the city border, their own filtration fields and thus decontaminate and remove all the refuse and other waste from their factories."[24] Trepov's argument reveals his conviction that the question of industrial pollution had been settled and that treatment through soil filtration was a universal solution. For him, tackling the problem of industrial pollution no longer depended on adequate technology and clear quality standards but solely on the willingness of industrialists to comply with the law.

These new local regulations were soon put into legal practice. Despite the intercession of Witte, who saw in the new policy a threat to the country's industrial development, any appeals for the increase of production were declined.[25] However, what looked like an excellent managerial and legal solution on paper turned out to be much more complicated in reality. The primary problem was the technological limitations of the sewage system. It soon became evident that not every type of liquid waste could be sent to the system, as certain effluents could clog, destroy, or corrode the sewer pipes or disrupt the biological processes on the fields. Furthermore, the sheer volume of wastewaters that Moscow's factories were producing by far exceeded the capacity of the sewage system. Already in 1903, the operation of filtration fields approached its limits, though half of the planned estates had not yet joined the system. These quantitative and qualitative limitations meant that, at the turn of the twentieth century, the sewage system could accept only a minor part of industrial waste produced in the city and was absolutely not ready to provide adequate material basis for the enforcement of the new local regulations on industrial pollution.

This put Moscow factories in a tight corner. As an example, the Giraud silk factory may illustrate how industrialists dealt with the situation. Together with other enterprises, the Giraud factory was denied authorization to increase production and reconstruct factory facilities, because provincial medical inspectors reported that its effluents polluted the Moskva River. In 1898, the factory owner, Claude Giraud, following the new recommendations and ready to bear all the incurring costs, requested permission to join the city's sewage system, but the municipality refused. Giraud appealed against this decision to the Commission on Industrial Discharge in the Moskva River, which sent him back to the municipality.[26] When, in 1899, Giraud again asked the municipality for permission to connect his factory to the sewers, municipal engineers responded that the volume of the factory's wastewater was too high for the capacity of the sewage system and that nothing could be done to resolve Giraud's problems with the law: "As for the police prohibition to discharge the dirty waters from the Giraud factory, other dyeing factories . . . are in the same situation, for example those of Kuznetsov, Gübner, Prokhorov, Zindel and many others, with much higher quantity of workers and huge amount of wastewater, for which the first line of the sewage system is not designed."[27]

Giraud did not give up and after several additional appeals eventually managed to find a technical solution with municipal engineers. In 1900, the silk factory was allowed to connect to the municipal sewers. This was, however, not a strategy that any Moscow factory could follow nor one that the municipality would encourage, given the limited capacity of the sewage system.

The impossibility of using only approved technology did not remove the pressure from the factories. Their expansion was prohibited and their operation allowed only temporarily and conditionally, while the law demanded that they be closed and their owners subjected to fines or even imprisonment. Although in practice these regulations hardly improved the water quality of Moscow's rivers, they did raise awareness of pollution among the industrialists and motivated them to look for ways to prevent or minimize it, even if only from fear of persecution.

In Search of Quality Standards

By the beginning of the twentieth century, the need for clearer definitions and standards of water quality had become evident to Russian legislators. In the 1900s, the Medical Council of the Ministry of Interior set out to elaborate norms for liquid industrial waste that was discharged in waterways. These standards, published in 1908 and revised in 1910, prescribed that the temperature of industrial waters should not exceed 30–40° Celsius, that they should not contain any poisonous compounds or pathogenic microbes and should not have any expressed color, smell, oil, fat films, or acidic or alkaline reactions. The last point of the regulations stated that industrial waste should not change the physical qualities and chemical composition of the river water "for the worse," so that it remained acceptable for the riverine flora and fauna.[28] Contemporary critics immediately pointed out that the last point indicated the general goal rather than a practical guideline of waste treatment. Such an imprecise and utopian formulation devalued the entire set of proposed quality standards and again meant that anyone could be prosecuted for breaking them.[29]

Although the new norms of the Medical Council could hardly work as clear guidelines of waste treatment, their publication entailed a wave of new court cases against factories responsible for water pollution in the Moscow industrial region. In 1910, the Moscow Exchange Committee (Moskovskii birzhevoi komitet), the nerve center of Moscow business elites composed of leading manufacturers and traders, received several appeals from industrialists of Central Russia, asking for protection against persecution on the grounds of polluting the waterways.[30] Such agitation among factory owners was due to the fact that, for the first time, the persecution for violating the water protection laws resulted not in a fine but in arrest and a prison sentence, which became apparent when the director of Tverskaia Manufactura was sentenced to several months in prison. The petitioners were generally skeptical about the risks of industrial pollution and the need of water protection. They complained about the high costs of waste treatment facilities, but their key concern reflected the problem that had long been recognized: "Nobody knows what degree of decontamination should be reached so that it is considered satisfactory."[31] The absence of clear quality standards hindered the evaluation of the waste treatment systems, as Provincial Medical Board, zemstvo, and municipal sanitary organization, factory inspectors, and police often came to mutually contradictory conclusions on which technologies could be considered acceptable. The lack of clear standards also meant that the logic behind the factory persecution was random and arbitrary.[32]

The members of the Moscow Exchange Committee, aware that they could not ask for the nonenforcement of the law, came up with a positive program that might potentially give industrialists legal grounds to argue against the arbitrary charges of the administration and police. They proposed initiating a large-scale research project to study industrial waste, its impact on public health and rivers, and the most adequate and efficient ways of its treatment. The results of this research would then form the basis for future regulations against pollution.[33]

This idea found support in the Ministry of Trade and Industry, the main ally of industrialists in the Russian government. The ministry managed to organized an interministerial conference on river pollution in December 1910. All the participants of the conference agreed that the time had come to revise pollution legislation and eventually favored establishing regional commissions that would work out local pollution norms and measures to abate it. Industrialists lobbied for such regional commissions in the hope that they would present "reasonable" antipollution demands and solutions adjusted to local types of production, specific pollutants, and the body of water in question.[34] Ministry officials hoped to use the findings of this research to develop the new countrywide law on industrial pollution that would finally settle the conflicts between industrialists and the police.

The first step in this direction was the creation of the Temporary Committee for Protection of the Waterways of the Moscow Industrial Region from Factory Waste (Vremennyi komitet po izyskaniiu mer k okhrane vodoemov Moskovskogo promyshlennogo raiona ot zagriazneniia stochnymi vodami i otbrosami fabrik i zavodov), which was approved by Emperor Nicholas II in 1911. The project involved several interested parties. It was supervised by high state officials from the Ministries of Trade and Industry, of the Interior, of Finance, of Communications, and representatives from the Factory Inspectorate as well as from local authorities. It was implemented by scientists—experts in chemistry, biology, bacteriology, engineering, and public health. Yet it was a private initiative organized and financed exclusively by private capital. The annual budget of the project (almost 95,000 rubles) was formed by the donations of 160 factories from the city and province of Moscow and neighboring regions. All the practical issues—outlays, contracts, instructions, office facilities, and scientific equipment—were arranged by the Moscow Exchange Committee.[35]

The involvement of industrialists in the elaboration of water legislation was an interesting but not a unique phenomenon. In Germany, mining, metallurgical, and chemical firms determined the action agenda of the water management authorities.[36] Similar precedents existed also within the Russian Empire. In 1902, the sugar producers of the southwestern provinces created and financed an expert committee that advised them on questions of waste treatment in their specific production and helped them settle conflicts with the local administration.[37]

In 1912–1917, this Waterways Protection Committee conducted extensive research on industrial discharge, its character, its relation to production, and the risks it posed. The reports of the Waterways Protection Committee reveal its members' particular interest in the impact of factory discharge on riverine environments in Central Russia. Committee experts investigated the effects of various industrial effluents on plankton and water plants, fish and crustaceans, insects and birds, revealing how deadly factories could be for the "river population" and for what we would now call biodiversity. For example, the "fish-tests" of the committee showed that untreated refuse from dye-work plants killed fish within thirty minutes.[38] The experts also studied existing filters and waste treatment systems in Russia and abroad and created test waste treatment plants in order to recommend the most efficient solutions. Their final recommendations are unknown, however, because the research of the committee was interrupted by the war and revolution.

Meanwhile, the government prepared the new antipollution legislation, which was submitted to the State Duma in December 1913. The new bill promised to resolve the question of pollution standards and obliged factory owners to install adequate antipollution equipment. Supervision of the execution of the regulations would be given to local public bodies (municipal and zemstvo governments). Although the proposed bill did not fully meet the demands of the industrialists, the special commission of the Association of Trade and Industry concluded that the new legislation, even with its defects, would still mean a relief from the existing administrative pressure and favored the adoption of the bill.[39] The onset of the First World War diverted the attention of the State Duma to other questions, however, and the antipollution legislation was never enacted.

The industrialists did not get any chance to benefit from the results of the research they were funding, as the Bolshevik revolution brought an end to private capital in the Russian industry. Yet the work of the Waterways Protection Committee outlived its sponsors as its goals and findings fitted well with the declared agenda of the young Soviet government. In 1919, on the basis of the Temporary Waterways Protection Committee, which was invented, organized, and financed by the imperial bourgeoisie, the Supreme Council of Popular Economy created the Central Committee for Water Protection, which would manage the water resources of the new proletarian state.[40]

Industrial waste in Russia emerged as a topic of public attention in the time of rapid industrialization and unprecedented interest and activity in the field of health and sanitation. Although medical professionals generally viewed industrial pollution as one of the factors affecting human morbidity, the fear of industrial waste had little direct influence on the shape of sanitary infrastructures in Moscow because exact links between specific pollutants and human health were rarely problematized and poorly understood. Sanitary reforms in Moscow, in their turn, had a major impact on policy toward the industrial pollution of waterways, both in the city and beyond.

It was public health fears during the cholera epidemic of 1892 that turned industrial pollution of waterways into a political issue. This did not happen, however, because industrial pollution was believed to be a cause of cholera (it was not) but, rather, almost by accident when the inspection to prevent contamination of waterways by organic wastes led to the discovery of the scale of illegal industrial discharge in Moscow. The cholera epidemic also set the pattern of administrative prosecution of sanitary violators that would persist for the next two decades and would be employed against industrialists in conflicts over water pollution.

It is interesting that the need to process domestic waste to prevent the spread of infectious disease was articulated before the need to process industrial waste. By the time the issue of industrial pollution in the city appeared on the agenda of the Moscow authorities in the early 1890s, the project of the sewage system with waste treatment facilities had long been under way. Urban sanitary infrastructure was seen as a pioneer in the field of waste processing and was taken as an example for industry, and the path dependencies created by municipal sewerage for human waste were awkwardly attached to industrial waste. The launching of the sewage system in 1898 as a part of the municipal public health reform validated the approach that waste should and could be neutralized and set higher demands for industrialists. However, at that time, understanding of the diversity and complexity of industrial effluents was very limited, while their composition, properties, and toxic compounds as well as their potential adverse impact on wastewater treatment processes and sewage farming had hardly been studied. This meant, however, that the view of the municipal sewage system and its analogues as the solution to the problem of industrial discharge was simplistic and inadequate. Still, this view did eventually lead to attempts to develop a more nuanced policy toward industrial waste and to look for clear quality standards.

Although Russian legislation had criminalized water pollution and unauthorized industrial discharge already before that, regulations on water protection grew stricter toward the beginning of the twentieth century, both on the normative level (higher fines and longer periods in prison as stipulated by law) and in practice (more active persecution of violators, harsher sentences). Yet the vagueness and imprecision of the legislation—acknowledged by industrialists, scientists, and imperial bureaucrats, liberals and conservatives alike—hindered and complicated the enforcement of water protection laws and at the same time opened the way for contingency and arbitrariness in the persecution of violators. The absence of clear quality standards was not, however, something peculiarly Russian but, rather, a common flaw of European legislation on industrial pollution at the turn of the twentieth century.

It is also possible to observe a gradual shift in perception of industrial pollution in Russia over this period. In the 1880s, the concern with industrial waste was part of the discussion on sanitation, with its anthropocentric approach, and the dangers of river pollution were seen mainly in threats to human health and to livestock raising (which was also reflected in legal norms). By the early twentieth century, the image of humans as the only and immediate beneficiaries of the fight for cleaner rivers became less obvious. Antipollution campaigns of the 1900s and 1910s had almost no reference to human health or direct economic interest but, rather, showed an attempt to protect river environments for their own sake, with human well-being as only one of the potential outcomes. Significantly, the necessity of (some) water protection appeared to be a matter of consensus, not of debate. Russian authorities, even those most supportive of industrial growth, never considered "sacrificing" rivers for the purposes of industry, although such ideas were voiced by factory owners and implemented in other countries. At the same time, the deteriorating environmental situation was increasingly connected to human action—in this case, of industrialists, who were openly declared not only guilty of pollution but also responsible for its prevention. It is plausible that recognition of industry's threats to nature was connected not only or even not primarily to environmental concerns but also to growing acknowledgment of the risks that factories posed to the social and political order of the Russian Empire, in particular, as hotbeds of socialist propaganda, workers' political mobilization, and the revolutionary movement.

The public discussion and legal prosecution of factory owners on the grounds of altering water quality did raise awareness of industrial pollution, they helped produce and accumulate knowledge about its risks and the practices of waste treatment. This knowledge, although not translated into policy during the imperial period, was passed on to the Soviet state. Yet the new government's declared commitment to abating river pollution and attempts to use the imperial experience in this sphere soon yielded to demands for industrial growth, and the aims and tasks of forced industrialization consigned water protection initiatives to oblivion.

PART III

Animal Bodies for Human Health?

Livestock and the Sanitary Project

CHAPTER 6

Feeding Moscow

Livestock Trade and Medicalization

"Meat should play a vital role in the nutrition of the human body as the food from which the muscles and other tissues and organs are formed. Meat is particularly important for a worker who constantly exhausts the strength of his muscles with hard labor; recuperation and formation of strong muscles for the working-class population is thus a matter of great significance."[1] Russian chemist Aleksandr Naumov, in his 1859 study of nutrients, voices one of the key dietary beliefs of the nineteenth century—meat makes muscles. Identified as a rich source of protein, meat was seen as the main source of physical strength and work productivity and as food indispensable for human health. The millennial practice of meat consumption was made intelligible in the works of chemists, physiologists, and hygienists, and in this new rationalized form became a part of political economy and a tool of managing the population.[2] Following their Western colleagues, imperial Russian scientists connected meat with strength, health, intellect, and resistance to epidemics and argued that more and better meat should be consumed.[3]

More meat, however, inevitably meant more livestock and more slaughtering. Supplying meat to the growing nineteenth-century cities involved profound reorganization of relations between city and country, when expanding hinterlands and new imperial peripheries were drawn into the orbit of the urban meat economy. Urban growth stimulated demand for fresh meat and new forms of large-scale livestock trade because, in big cities, huge quantities of perishable fresh meat could be sold within a short time. Advances in transportation and communication and new technologies of slaughtering and refrigeration enabled the expansion of the local supply chains into national and international markets and transformed vast expanses of land for commercial livestock production.[4]

The Russian Empire is often thought of as a grain country, but it was also home to one of the world's largest livestock populations and a site of long-distance livestock trade.[5] The expansion of the Russian Empire to the south and east in the eighteenth and nineteenth centuries and the colonization of the frontier regions north of the Black Sea and the Caspian Sea allowed for the rise of commercial animal husbandry and the growing spatial separation of livestock raising and meat consumption. As a result, hundreds of thousands of animals, especially cattle, had to be regularly moved from the Eurasian steppes to the slaughterhouses of Central Russia. If the American cattle drives of the second half of the nineteenth century are among the most famous and most romantic frontier icons, Russian cattle drives are far less known. Similar to the American cowboys moving herds of animals from the prairies of the Midwest to the meatpacking plants in Chicago, their less iconic Russian colleagues drove cattle for even greater distances across the Romanov empire, bringing the resources of the steppe to the table of the Russian urban public.[6]

Moscow was Russia's largest center of cattle trade, meat production, and consumption. Its role as a meat hub depended on complex connections with its distant hinterland and imperial peripheries, which transformed regional environments, economies, and the relations between humans and animals. The study of Russian meat production reveals that, in the second half of the nineteenth century, this process was profoundly entangled with the rise of public health, human and veterinary medicine, and that medical concerns played a major role in shaping commercial livestock raising, meat industry, and the animal experience.

Moscow and Its Meat Hinterland

When one looks at the geographies of Moscow's meat economy, the most surprising aspect perhaps is how far animals had to travel to end up on the plates of Muscovites. Even though Moscow was surrounded by agricultural provinces, that region played a minor role in the city's meat supply. Unlike in many cities of continental Europe, where pork was the prevailing meat, the market in late imperial Moscow was strongly dominated by beef

TABLE 6.1. **The Origin of Cattle in the Most Important Regions of the Moscow Market**

1885 (Head of Cattle)		1903 (Head of Cattle)	
Total 216,871		Total 258,773	
Six most important regions		Six most important regions	
Don Cossack region	72,015	Don Cossack region	68,558
Kharkov province	27,716	Voronezh province	42,180
Voronezh province	22,005	Saratov province	37,670
Kuban region	18,196	Kharkov province	36,213
Ekaterinoslav province	17,258	Samara province	17,379
Poltava province	13,218	Kuban region	10,845

Source: *Otchet veterinarnogo otdeleniia MVD za 1885 g.* (Saint Petersburg, 1890); *Otchet veterinarnogo upravleniia MVD za 1903 g.* (Saint Petersburg, 1906).

and was dependent on the long-distance cattle trade. In 1890, for example, 154,000 head of cattle, 47,000 calves, but just 23,500 pigs and piglets were slaughtered in Moscow. Only calves came from Moscow's immediate hinterland. The main hog-raising regions were several hundred kilometers southeast of Moscow in the provinces of Ryazan', Tambov, Voronezh, and Penza. Meat cattle, however, were coming to Moscow from much further away.[7]

In the densely populated provinces of Central Russia, with their land scarcity and shortage of pastures and meadows, cattle were kept primarily for manure and milk. These were mostly small, short, skinny cows. Despite the proximity to the markets of Moscow and other industrial centers, commercial cattle raising for meat was uncommon in this region because of the lack of suitable fodder. Peasant households did not have enough grass and especially hay to feed the animals during the long winter months. Livestock malnutrition was a real problem, and the cattle were valued for their endurance and ability to give high milk yields despite a poor diet. With available fodder, the meat quality and the market price of adult bulls was so low that it was more profitable to sell and slaughter animals as calves than to raise them for meat.[8]

However, what was scarce in Central Russia could be found in abundance in the vast grasslands in the southern and southeastern part of the Russian Empire. The cattle slaughtered and consumed in Moscow came from the steppes of the Dnieper, the Don, and the lower Volga Rivers, the North Caucasus and Central Eurasia (see table 6.1).

This region, which includes the territories of present-day Ukraine, Russia, and Kazakhstan, had depended on livestock for millenia. The economy of many nomadic populations that inhabited the steppe—the Skythians, the Khazars, the Pechenegs, the Mongols, the Nogai, the Kalmyks—was based on grazing and trading livestock, primarily sheep and horses. The entire treeless landscape of this region, no matter how wild and pristine it seemed to Russian imperial observers, has been formed by many centuries of nomadic pastoralism and grazing. The Russian conquest of this region started in the sixteenth century, when the tsar Ivan the Terrible defeated the Tatar khanates of Kazan and Astrakhan, which gave Muscovy control of the Volga River and its delta in the Caspian Sea. In the next two centuries, however, Russian expansion to the Sea of Azov and the Black Sea was hindered by the Crimean Tatar Khanate, a vassal state of the Ottoman sultan, that controlled not only the Crimean Peninsula but also the steppes between the Dnieper and the Don. Along its southern and southeastern frontier, Russia faced a constant armed conflict, with small-scale raids by steppe nomads as the most common frontier encounter, which subverted the efforts to settle the region. It was not until the end of the eighteenth century that Russia secured its position in the European steppes, after decisive victories over the Ottoman Empire and the annexation of the Crimean Khanate in 1783. The process was finalized with the Caucasian War and Russian expansion into Central Asia in the nineteenth century, bringing the largest part of the Eurasian steppes, from Bessarabia in the west to the Chinese border in the east, under the control of the Russian Empire.[9]

The conquest of the steppe unleashed the processes of colonization and settlement. The population of the European steppe increased from 380,000 in 1719 to over 8 million in 1850 and 25 million on the eve of the First World War.[10] The removal of the military threat opened the way for agriculture and livestock raising on these vast, fertile, and at first sparsely populated territories. In fact, agriculture took some time to spread, especially in the southeastern steppe region. In particular among the Cossacks, livestock raising remained the main economic activity until the second half of the nineteenth century. In the steppe, large herds of livestock could be maintained with comparatively little labor, which was usually in short supply among the Cossacks because of the military service (while tending to livestock was often outsourced to women and children).[11]

At first, new settlers in the steppe continued raising horses and sheep, as the nomads had done before them. Steppe horses were in high demand for military purposes, both among the Cossacks in the southern provinces and in other regions of the Russian Empire, whereas sheep were the main source of meat for local consumption. In the early nineteenth century, however, sheep started to give way to cattle. The rise of cattle was linked to the changing land use in the steppe, to the increased demand for meat in the growing urban centers, and to the spread of arable farming.

The traditional Russian horse-drawn plough could not cut through the steppe grasses and soil. Instead, the main agricultural instrument used in the steppe was the heavy Ukrainian plough. That plough was pulled by several oxen, making them the most widespread working animals in the steppe region. Oxen were cheaper to feed than horses and were better suited to work on the steppe soils that had not been plowed for many years. Unlike the small, skinny dairy cows of Central and Northern Russia, these were huge animals, valued for their muscles. When those oxen could no longer be used as workforce, they were given some time to fatten and then were sold for meat. In addition, there were cattle raised directly for meat. Although the two types of cattle raising often coexisted within the same region, the second type was prevalent in the eastern part of the European steppes and the steppes of Central Eurasia, where arable farming was less common.[12]

By the middle of the nineteenth century, mass cattle drives from the steppe became widespread, forcing the government to regulate this trade and restrict it to specific cattle-driving trails (*skotoprogonnye trakty*). Those trails were delineated stretches of land, usually between twenty and one thousand meters in width. The width of the trail meant it was also used to graze the cattle along the way. In addition, cattle were grazed and watered on the private lands bordering the trail, either for free or for a fee, and state regulations forbade the monopolization and enclosure of such pastures. Those measures were explicitly aimed at facilitating cattle trade so as to ensure a stable meat supply to Moscow and Saint Petersburg and to keep meat prices in the cities low, even at the expense of landowners and local communities whose lands had to stay open for commercial cattle. Until the 1850s, grazing along the trail was the main way of making the animals gain weight. Later, however, with the rise of arable farming, the growing population density, and the legal restriction of free grazing in 1867, this practice was supplemented by feedlots, either at a special pasture (*polevshchina*) or principally in Ukraine, at distilleries or breweries, as the by-products of alcohol production (*barda*) were believed to be an excellent fodder for meat bulls.[13]

There were several main directions of cattle drives across the Russian Empire. By far the longest routes led from Siberia and Central Asia. They started with the herds bought from the nomads ("Kirgiz") in Akmolinsk,

Semirechinsk, Turgai, and Tobolsk regions (present-day Kazakhstan) and the herds from the Orenburg province and the Ural Cossacks. After crossing the Ural Mountains, those herds were joined by the cattle from Samara, Ufa, Saratov, and Astrakhan, and then moved to the markets of Kazan', Nizhnii Novgorod, and Moscow. Some of the cattle on these routes had to walk for more than three thousand kilometers before reaching their destination. Another group of trails led to Moscow from the northern foothills of the Caucasus and the steppes between the Caspian, the Black Sea, and the Azov, with the herds from Stavropol and Kuban, from the Don Cossack region and then Tambov and Voronezh. In the second half of the nineteenth century, the Don Cossack region remained the main source of cattle for the markets of Moscow and Saint Petersburg. The third group of cattle-driving routes connected Moscow to the steppes on the left bank of the Dnieper, in the Kharkov, Ekaterinoslav, and Poltava provinces through Belgorod and Kursk. Finally, Ukrainian cattle from the steppes between the Dnieper and the Dniester Rivers went in several directions: northeast to Moscow, north to Gomel and Nezhin (present-day Belarus), west toward Warsaw, and south to Odessa. The cattle on the market in Saint Petersburg came largely from the steppe through Moscow. Unlike Moscow, however, Saint Petersburg also consumed considerable numbers of cattle from the neighboring regions of the empire, for example, from the Pskov province, Estland (today northern Estonia), or Finland.[14]

The colonization of the steppe frontier, the rapid increase of population, and the change of land use that had initially contributed to the rise of cattle raising in the European steppe would eventually lead to its decline. As more and more land was plowed up and converted to grain fields, fewer pastures were left for the cattle to graze. By the beginning of the twentieth century, arable farming gradually replaced livestock raising in the steppes of Ukraine and southern Russia, pushing it further east. At the same time, the role of the Caspian, Ural, and Siberian steppes in the commercial cattle production for the Moscow market increased. In 1903, the Ekaterinoslav and Poltava provinces, which had been the key suppliers of meat animals in the 1880s, together sent less cattle to Moscow than the Orenburg province alone.[15]

This long-distance cattle trade relied on a complex system of social relations, with many stages and intermediaries between the peasants, Cossacks, or steppe nomads on the one hand and the meat consumers on the other. The key figure in this system was a *prasol* or a *skotopromyshlennik* (literally, a "livestock industrialist")—that is, a cattle drover. Prasols, who themselves usually came from Central Russian provinces, bought cattle at the steppe fairs and formed the herds. To speed up the purchase they employed so-called runners who ran around the fair to find animals of the right breed and shape. Depending on the quality of the cattle, the herds could be sent to the feedlot for several months before starting their way to the urban markets.[16]

Throughout most of the nineteenth century, cattle moved from the steppe pasture to the meat market on foot, walking on average between 10 and 30 kilometers a day, although the speed depended on the condition of the animals and the route, available fodder, and prices at the destination market. A herd of 200 head of cattle required a team of five cowboys, a cook, and a prasol or his representative. In the 1870s, a trip on the Don Trail from Rostov to Moscow (1,000 kilometers) took 60 days; a trip on the Siberian Trail from Orenburg to Saint Petersburg (2,000 kilometers) required 114 days. Most of the drives took place in the summer and fall months so that animals could graze on the way, as feeding cattle with hay doubled the cost of the drive.[17]

In the city, the cattle were sold once again and went from "livestock industrialists" to "meat industrialists" (*miasopromyshlenniki*) or "bull killers" (*bykoboitsy*). This was a small and wealthy group of wholesale businessmen who, as their professional name suggests, made profits from turning living animals into meat. They sent the cattle to slaughter, where a herd of 200 bulls would transform into 60 tons of fresh beef, which would then be sold to retail butchers.[18] Before appearing on the plates of Muscovites, animal bodies, living or dead, changed hands at least four times, when each new transaction meant a new degree of alienation and commodification.

The rise of the long-distance livestock trade between the steppe regions and the urban markets of the Russian Empire was not just a socioeconomic reality. It was itself a cause and a result of the dramatic environmental transformation connected to the settlement of the steppe. The trade also had profound consequences for the complex multispecies relations in Russian food systems, agriculture, and animal husbandry, as the herds of livestock moving for hundreds and thousands of kilometers were an ideal vehicle for many microorganisms that lived in and on them. The constant migration of thousands of animals across the empire not only connected the peasant or Indigenous producer with the urban consumer but also linked disease ecologies of the grasslands and the city as well as of those many spaces that the livestock crossed through along the way. Herds of commercial livestock, where animals from various localities and households came together in large numbers, were perfect reservoirs of disease. As animals passed through fairs, feedlots, pastures, and watering places, infections were transmitted not only to other members of the herd but to the local livestock—and sometimes the human population as well. As it soon became apparent to Russian officials, economy, environment, animal health, and human health were all tightly linked.

Disease, Medicine, and Livestock Trade

The nineteenth century marked the institutionalization and professionalization of veterinary medicine, both in Russia and worldwide. The growth and internationalization of the livestock and meat trade led to the globalization of infectious animal disease and increased the risks to livestock populations and the human economy. The new understanding of the links between human and animal health—resulting from bacteriological discoveries, acceptance of laboratory medicine, and the study of the mechanisms of transmission and immunity—meant that both the veterinary profession and the subject it was dealing with moved from the margins to the center of public attention. In the 1870s and 1880s, the Russian Empire experienced a transition to what Joanna Swabe calls the formal veterinary regime, marked by the formalization of veterinary education, the professionalization of veterinary medicine, and the increasing role of the state in dealing with animal disease.[19]

If there was one disease that shaped veterinary medicine, livestock trade, and human-animal relations in late imperial Russia, it was rinderpest (cattle plague). Although not dangerous to humans, this viral disease was extremely contagious and usually fatal to cattle, disrupting the agricultural economy and undermining the well-being of the population in affected regions. The disease was enzootic in the Eurasian steppes, where local cattle usually displayed only mild symptoms but could serve as a vehicle of infection.[20]

Outbreaks of cattle plague had been known in Russia at least since the eighteenth century, but the rise of the market-oriented livestock economy in the steppe frontiers and the expansion of the cattle trade geography facilitated the spread of disease, both in Russia and, through its trade, transportation, and warfare, further west in Europe. In 1865–1867, Britain was hit by a devastating outbreak of cattle plague that killed 420,000 head of cattle; it was introduced by the imported livestock from Eastern Europe, most likely from the Russian Empire.[21] In 1871–1873, cattle plague claimed almost 700,000 head of cattle in Russia, making it an issue of national economic importance.[22] Although the virus causing the disease and the exact mechanisms of transmission were not known, it was accepted that cattle plague was infectious and that herds of commercial livestock from the steppe played a key role in spreading it.[23]

To cope with this problem, in the 1860s and 1870s the Russian government passed several regulations that would transform livestock trade, food systems, and human-animal relations more generally. In 1869, the control over animal health in commercial herds was transferred from the police to the newly established veterinary service on the livestock trails, financed by a special fee paid by cattle drovers.[24] In 1870, new regulations against cattle plague were introduced in the Kingdom of Poland, the region of the Russian Empire that played a key role in the country's (generally moderate) livestock exports. Those regulations prescribed compulsory killing of plague-infected or suspicious cattle. Similar practice of stamping out and slaughtering infected animals existed in many European countries, where it had been widely applied to prevent the introduction of disease by imported cattle, including those from Russia.[25] In 1877, the killing of plague-infected animals was made compulsory in commercial herds.[26] Two years later, in 1879, this regulation was extended to the rural livestock in the core Russian provinces as well, when the Ministry of the Interior obliged zemstvos to exterminate plague-infected animals, with compensation to be provided to their peasant owners.[27] Compensation was a crucial point of those regulations. This measure was meant to prevent peasants from hiding the sick animals or selling their products. It also encouraged them to look for the symptoms of infection and report it early, before the animals actually succumbed to cattle plague—the compensation was provided only for the animals that were killed and not for those that died themselves.[28]

These policies were a remarkable and remarkably effective example of more-than-human biopolitics. By the 1890s, the rinderpest epizootic showed clear signs of decline, both in the numbers of casualties and in the regions where it spread. In 1875–1879, more than 2 million head of cattle died of rinderpest; in 1890–1894, only 95,000, and a majority of those were killed as a veterinary measure. If in 1881 the epizootic affected forty-eight provinces of European Russia, ten years later the disease was present only in the provinces of Samara, Tobolsk, and Ufa. This was a stunning success.[29]

Another outcome of Russia's effort to fight epizootics was the transformation of animal movement across the country. Starting in 1884, all livestock going to the markets of Moscow or Saint Petersburg were obliged to be transported by railways to minimize the contact between the commercial herds and local animals. This measure eventually meant the end of the mass cattle drives through Central Russia. Cowboys now had to deliver their cattle only to the nearest station on the expanding railway network, and their trade was gradually pushed out of European Russia to the southeastern borderlands, the Ural and Siberian steppes.[30]

The change to animal transportation by railways was, of course, a process known to other countries as well. In his famous account of the American meat trade, Bill Cronon explains this transition from "meat on the hoof" to "meat in a railroad car" by the logic of capital, where cattle owners found it cheaper and more efficient to ship cattle by rail rather than to drive them overland.[31] I would argue, however, that in the Russian case this transition did not happen as a response to the pressures and demands of the market, at least not directly, but was imposed from above and motivated by medical concerns. Essentially, mandatory railway transportation of livestock was a sanitary measure, and the railway stations were integrated into the system of veterinary control. Railway transportation of livestock should be seen together with growing medical control over livestock and compulsory killing of sick animals and was a part of a broader effort against animal disease.

In fact, Russian cattle businessmen strongly opposed the introduction of railway transportation. They complained about the negative consequences this measure had on the livestock trade and on the quality and availability of meat in Russian metropolises. If, in the United States (at least in Cronon's interpretation), railroad transportation served to protect the value of cattle by preventing weight loss on the way to market and streamlining the delivery process, in Russia it seemed to have none of these effects. Russian cattle owners often explicitly compared their business to that in the United States, offering several explanations as to why this economic rationale did not work in the Russian context.

One reason was the costs of the overland drive versus railway transportation. In Russia, human labor was cheap and the grazing along the trail almost free, so railway transportation with existing tariffs meant a two- or threefold increase of associated delivery costs. Although the new technology made delivery somewhat faster, cattle owners felt that overland drives gave them more control over the process, when one could always adjust the arrival of the cattle to the market situation, whereas with the railroad, the shipment was bound by train capacity and schedule.

The more important argument, however, was the animal experience of railway transportation. In comparison to the United States, Russian trains were slower and had smaller and less convenient stock cars. They were significantly louder and joltier because of their biaxial design and also lacked adequate facilities for feeding and watering the cattle. As a result, in Russia, the railway trip was arguably more stressful and damaging for the animals, and they lost tremendous amounts of weight on the way. Considering that the long overland drives in Russia had not been connected to weight loss but, on the contrary, were a method of fattening the animals on the way to the market, it is not surprising that cattle businessmen did not see railway transportation as being favorable to their trade and accepted it only as a necessary medical measure to prevent epizootics.[32]

Livestock and Community Medicine

The new medical regulation of the livestock trade was a watershed in the history of both veterinary medicine and human-animal relations in the Russian Empire. For the first time, Russian livestock was exposed to the medical gaze on a large scale. This regulation prompted the institutionalization of veterinary medicine in Russia, the accumulation of veterinary knowledge, and the creation of medical structures that would persist and expand after the rinderpest itself was brought under control. Within several decades, Russian veterinarians not only managed to gain the status of a modern profession but emerged as experts of livestock animals, as the main mediators of human-animal relations, and the judges of how these animals and their bodies should be treated.

Veterinary medicine quickly became integrated with community medicine structures. As with physicians, the growing zemstvo and municipal organizations provided new employment opportunities for veterinarians and enhanced professional exchange, collaboration, and identity building. In the 1860s and 1870s, most zemstvos had not employed any regular veterinary personnel. To comply with the new regulations on rinderpest, many had to hire veterinarians, which brought medical expertise in animal health to local communities. Zemstvo sanitary congresses have often been credited for the emergence of Russian community medicine. Russian veterinarians also used this format to share their experiences, articulate their concerns to the local governments, and collectively work out consistent policies. The first provincial zemstvo sanitary congress took place in 1871, the first veterinary congress three years later, and in the following decades this practice gained wide popularity. By 1902, veterinary medicine had a section within the Pirogov Society of Russian Physicians, the most influential and politically vocal medical organization in the country and the mouthpiece of community medicine.[33]

The quick retreat of rinderpest allowed the growing veterinary organizations to direct their attention and resources to other animal diseases such as anthrax, contagious bovine pleuropneumonia, glanders, foot-and-mouth disease, and rabies. As some of these diseases were zoonotic, this expansion of focus meant that veterinary medicine was increasingly becoming a discipline and profession that contributed to protecting human health and explored its complex connections with lives of other species. If the fight against rinderpest concentrated on the medical police measures of stamping out, quarantining, and extermination, the decline of this epizootic allowed for the diversification of medical strategies used by veterinarians. The diversification was also helped by bacteriological discoveries and international progress in disease diagnostics and prevention and included, for example, the wide vaccination of Russian livestock against anthrax, a disease that can potentially be lethal to humans, and to a lesser extent swine erysipelas, starting from the 1880s.[34]

On the other hand, the rise of veterinary institutions to fight rinderpest, combined with the eventual quick decline of this disease, allowed for the development of therapeutic medicine for animals, when an encounter with a human medical professional promised not only quarantining or extermination but therapy and cure. In the 1890s, zemstvos started opening facilities for the medical treatment of animals. As the board members of the Moscow provincial zemstvo reported in 1894, such "animal clinics" not only served their immediate purpose of healing livestock but also helped detect infectious disease and promoted trust and better understanding between peasants and veterinarians. By the early 1900s, these clinics were treating more than a million animal patients annually.[35]

Integration of veterinary medicine with the community medicine was happening not only in the countryside but in the cities too, and Moscow was a striking example of it. Throughout the 1880s and 1890s, various aspects of human-animal relations in Moscow were put under municipal veterinary control. The largest and most expensive innovation was the slaughterhouse reform and the construction of the Moscow municipal abattoir in 1886–1888 (discussed in the following two chapters).

In 1887, the Moscow City Council passed regulations on glanders, a contagious bacterial disease of horses that can also be contracted by dogs, cats, and pigs, and which can be lethal to humans. The discovery of the glanders bacillus was one of the early successes of bacteriology. It was isolated by Friedrich Loeffler and Wilhelm Schütz in 1882, the same year that Robert Koch isolated the bacillus of tuberculosis. In 1891, inspired by Koch's discovery of tuberculin, Christopher Helmann from the Institute for Experimental Medicine in Saint Petersburg and Otto Kalning from the University of Dorpat developed a mallein test that became an important diagnostic tool for glanders (unfortunately, both Kalning and Helmann died of glanders soon after that discovery).[36]

The new municipal regulations against glanders in Moscow, modeled on the national regulation on rinderpest, introduced compulsory killing of diseased animals with a generous compensation to their owners from the city budget. In 1888, the city conducted a comprehensive horse inspection, examining 26,000 animals and revealing 74 cases of glanders. In 1894, the municipality introduced registration of the city cattle (numbering 7,000 that year) and passed regulations against bovine pleuropneumonia, anthrax, rabies, scabies, and foot-and-mouth disease. Those regulations, also based on the rinderpest model, made urban milk cows, dogs, and cats the target of medical inspection and control, stamping out, and extermination.[37] To enforce the new regulations against animal disease, the city government established a municipal veterinary organization in 1889–1891.[38]

The city also developed therapeutic veterinary medicine. Interestingly, access to municipal health services was given to humans and (nonhuman) animals almost at the same time and in a similar format. In Moscow, the municipal outpatient clinics for humans opened in 1886 and for animals in 1889. The clinics were managed by the same unit of the municipal board, had the same name (*ambulatoriia*), and followed the similar principles of district organization, case registration, and free services. In 1894, Moscow had four such animal ambulatoriia, which served 8,762 individual animal patients that year.[39]

Importantly, veterinary inspection and the control of animal bodies were described and classified in the city budget as a "sanitary measure" and were listed and financed together with the work of sanitary doctors, street cleaning, waste removal, the inspection of prostitutes at the women's outpatient clinic, and actions taken against epidemics. From the very beginning, this sanitary measure embraced new bacteriological knowledge and the tools it offered—for example, the mallein test for glanders that was used in Moscow from 1892 onward. It allowed for a quicker and more efficient diagnosis, brought advances in bacteriology to the sanitary reform, and served as a justification for animal expropriation and slaughter.

The sanitary measures against epizootics had profound economic, spatial, sanitary, environmental, and cultural consequences. The redirection of the livestock trade to shorter overland routes and railroad stations hid it from the majority of urbanites and obscured the relations between the faraway steppe and the market in Moscow. It led to the further commodification of the animal body: if the overland trade still depended on the cattle's ability to walk and graze during their journey to the market, with railroad transportation animals were packed together into a freight car without water or food and were treated like lifeless cargo already at the beginning of the trip. As the delivery by rail eliminated grazing along the way and was associated with significant weight loss in transported animals, it increased the trade's dependence on a feedlot. With free land in the steppe becoming scarce, the large ranching fields with free grazing gave way to feedlots, where animals were squeezed in a limited space, often intentionally without the ability to move, and where their entire existence was subordinated to the ultimate goal of gaining weight. This also led to the development of a special diet of processed fodder and the entire science of animal fattening, tasked with ensuring that the weight gain happened in the right form and place, which affected not only the shape and health of the animals but also the flavor of the meat itself.

Within a period of several decades in the second half of the nineteenth century, Russia experienced a dramatic change in how livestock animals were raised, fed, moved, and controlled, and medical concerns and medical professionals played a major role in this process. This medicalization went together with the rise of veterinary research and veterinary statistics, which contributed to the inspectability of human-animal relations. Patterns of animal disease and animal lives more broadly were described, calculated, mapped, and rationalized through the language of medical science. The structures that monitored and controlled animal bodies were integrated with the system of community medicine and became part of urban and rural sanitary organizations. As the next chapter will demonstrate, the transition from a living animal to edible meat was also transformed beyond recognition.

The process of medicalization and the effective biopolitical interventions were celebrated by veterinarians and public administrators, impressed by the sharp decline of cattle plague—an exceptional case of such spectacular success in combating disease in the Russian Empire. Although the target was not the human but the animal population, this case proved that humans have the power to control disease and that coordinated efforts of experts, the government, and the public could lead to the eradication of a dangerous and widespread infection. Both medical professionals and civil reformers viewed medicalization as a sign of progress, when animal health, disease, life, and death could be monitored and managed to benefit human well-being.

CHAPTER 7

The Killing Factory

In the nineteenth century, many cities saw a remarkable transformation in how meat was produced. The concentration of population in the urban centers pressed the traditional art of butchering to increase in both scale and speed, while the developing sanitary and medical sciences pressed for stricter control over the slaughter. The modern preoccupation with civility, sanitation, and order demanded a dissociation between healthy and nutritious meat and the act of killing that it implied. The transformation of the slaughterhouse reflected a profound shift in European sensibilities. Previously, animal death had been a daily experience in urban life. The herds of livestock intended for slaughter regularly passed through the city streets, and some were killed right behind the butchers' shops. Modern sensibilities and the new hygiene regimes demanded that death, blood, and physical violence as well as disease, foul odors, and pollution be removed from the increasingly ordered and "civil" city. The public abattoirs, together with other products of modernity such as the prison, the clinic, and the sewerage systems, were the mechanisms of this transition.[1]

As the anthropologist Noelie Vialles puts it in her study of French abattoirs, "slaughtering was required to be industrial, that is to say large scale and anonymous; it must be non-violent (ideally: painless); and it must be invisible (ideally: non-existent). It must be as if it were not."[2] The slaughter was exiled to the outskirts, enclosed and confined within the walls of the new institution; it had to be marginalized, hidden, excluded from everyday life, and turned into a "no place." Even the euphemism of its name—*abattoir* instead of the French *tuerie* or the English "slaughterhouse"—was meant to disguise the violence of its purpose.[3] In the last twenty years, cultural historians have taken this formula of "a place that was no place" to explore the meaning of the slaughtering reform in other cities across the Western world, emphasizing the intention for anonymity, invisibility, and dissimulation embedded in the projects of the modern abattoirs.[4]

The public abattoir in Moscow was a case that in some respects contrasted sharply with the pattern outlined by Vialles. Constructed in 1886–1888, the centralized municipal abattoir in Moscow, as in many other European cities, came to replace small private facilities and to transform slaughter into a more efficient, hygienic, and controlled process. Yet, instead of becoming a "no place," it faced a rather different prospect: to turn into one of the city's most recognized infrastructural projects and, in the words of a contemporary, to "join the rows of the institutions that constitute the city pride such as museums, art galleries, universities, and the like."[5] Looking at the construction and operation of the slaughterhouse in the context of Moscow sanitary reforms and the medicalization of human-animal relations, I will explore how and why the place of industrial slaughter avoided marginalization and appeared as one of Moscow's landmarks.

Meat Production and Public Health

Reorganization of meat supply and slaughterhouse reform emerged as an important feature of urban modernity after the Napoleonic era, and during the second half of the nineteenth century new abattoirs and stockyards were built all over the Western world.[6] In Russia, the necessity of such reform had been discussed since the 1840s, but the first public abattoir appeared only in 1882 in Saint Petersburg. This example was soon followed by other cities of the empire, including Moscow.[7]

Moscow's path to the public slaughterhouse was long and winding. In the early 1860s, the Moscow City Council discussed the reorganization of existing private slaughterhouses in view of their dirt, stench, and "unsatisfactory condition" and concluded that the only way to improve the situation was to open municipal abattoirs. For these purposes, in 1866, the city bought a plot of land south of Moscow, in the area of the Serpukhov Gate, which was then the main hub for livestock and the destination of the cattle trails.[8] This initiative, however, was soon lost in municipal discussions and then abandoned until 1885 when the city council returned to the project.[9]

In the meantime, the city and the entire country changed dramatically as a result of the Great Reforms of the 1860s and 1870s. The medical agenda, too, was experiencing the paradigmatic shift with the rise of veterinary medicine, bacteriological discoveries, establishment of the laboratory, better understanding of disease transmission, and the links between the animal and human health. The introduction of mandatory railway transportation of livestock brought an important change to the spatial morphology of Moscow's slaughtering arrangements. The old livestock trails lost their significance and so did the plot of land to the south of Moscow that the city had bought for its intended slaughterhouse. Instead, the herds of animals were now coming to the terminals of the Kursk and Ryazan railways, located on the east of the city. This pressed for the relocation of the slaughterhouse eastward to be close to the stations. Arguments in favor of such a relocation were framed mainly in sanitary and medical terms: this would not only spare the city streets the inconveniences of cattle drives but would also prevent the spread of epizootics from the imported animals to the local milk cattle.[10] The centralization of arrival opened the way for the centralization of veterinary inspection and slaughtering and promised to turn the trip from the steppe pasture to the Moscow meat market into a more controlled but less visible process than ever before.

If rinderpest stimulated the institutionalization of veterinary medicine and animal inspection, it was trichinosis, a disease associated with swine, that connected meat production to the scientific laboratory. Caused by a parasite roundworm, *Trichinella spiralis*, this disease was discovered by James Paget in 1835. In the second half of the century, Rudolf Virchow and Friedrich Albert von Zenker described the life cycle of *Trichinella* and revealed that humans were at risk of contracting the disease through eating pork.[11]

In Russia, the first detailed description of this disease appeared in 1862 on the basis of reports from abroad. However, when the pathologist Mikhail Rudnev, one of Virchow's students, discovered the case of trichinosis in a dissected female corpse in Saint Petersburg in 1865, it became clear that the disease was present in Russia as well.[12] In following years, his colleagues reported incidents of trichinosis from Moscow, Saratov, Kharkov, Riga, and other cities. Microscopic examination of meat was seen as the only way to ensure its safety; otherwise, the experts advised, eating pork should be completely avoided.[13]

Although pork consumption in Moscow was relatively low and cases of trichinosis very rare, the disease managed to attract public attention and received significant press coverage. One possible explanation of this could be the influence of German scientists and policies. In Germany, where pork was the most important meat, this was a much more real health problem, with higher incidence and mortality. There were several notorious outbreaks of trichinosis in Germany in the 1860s–1880s, with the deadliest one occurring in a village called Hedersleben in 1865, when 300 of its 2,000 residents fell ill and 100 died. These outbreaks sparked interest in a wider scientific community and eventually became the impetus for the slaughterhouse reform in Germany.[14] Considering the tight scientific connections between Germany and Russia and the international attention that German medicine was enjoying at the time, it is plausible that the Russian public interest in trichinosis was an echo of a German discourse.

Another possible explanation of that interest could be more broadly cultural. The scope of social responses to a disease are rarely determined by its incidence alone. Other factors are usually at play: its newness or exoticism, its power to invoke fear and repulsion, its ability to reveal other societal problems or to fuel relevant political debates. Trichinosis definitely had the potential of capturing the human imagination. Its association with pigs and tiny worms and its severe, painful, and visually repelling symptoms, including swelling, paralysis, and suffocation, made it both disgusting and frightening. Even more terrifying, as Dorothee Brantz suggests in her historical study of this disease in Germany, was the awareness that tasty food could kill and do so without any obvious indication of danger.[15] The usual mechanisms of sensory vigilance that butchers and consumers had used to avoid rotten meat—visual examination, smelling, and tasting—were powerless against trichines, which were both tasteless and invisible without a microscope.

In 1876, the Medical Council of the Ministry of the Interior discussed the questions of trichinosis in Russia and concluded that the meat of trichined animals was to be prohibited from sale, forage, or any other use and was to be subject to immediate destruction. To put this ban in practice, the ministry recommended introducing microscopic examination as an important step in pork production.[16] The centralized abattoir equipped with laboratories and an adequate system of veterinary inspection therefore came to be seen as a mechanism to ensure the safety of meat and livestock. This "veterinary turn," as the members of the Moscow Slaughterhouse Commission (Kommissiia po ustroistvu boien) acknowledged, should be reckoned with when devising a project for the enterprise: "Previously, the only demand for the improved slaughterhouse was that it was kept clean and did not produce any foul odors. Now this is not enough. From the veterinary side it is required that the slaughterhouse helps to combat rinderpest, raging in Russia. From the sanitary side it is considered necessary that the slaughterhouse serves as a controlling point for the quality of meat to prevent the sale of meat from sick animals."[17]

Finally, not the least part in the story belongs to the example of Saint Petersburg. The traditional rivalry between the "two capitals" of the empire entered a new phase when, in the second half of the nineteenth century, Moscow emerged as a center for the growing Russian bourgeoisie—as opposed to the socially and economically "westernized" Saint Petersburg.[18] In 1882, Saint Petersburg built the centralized municipal slaughterhouse, which not only proved that such an institution could function successfully in Russia and provided an illustration of how it could be achieved but also included the questions of city pride in the agenda of the slaughtering reform.

In May 1885, commission members prepared a preliminary plan for the new Moscow abattoir. They proposed moving it to another location southeast of the city and connecting it with a special branch railway to the main routes of cattle transportation. In addition to the infrastructural advantages, this location, considering Moscow's compass rose with prevailing western winds, spared the city from the odors of the slaughter. The complex was also supposed to include a stockyard, storage facilities, and factories to process blood and tallow.[19]

Although the preliminary project was generally designed according to the model of Saint Petersburg, there was a crucial difference. The abundance of water in Saint Petersburg and the proximity of the Baltic coast offered the city an easy solution to the question of slaughterhouse sewage, which was simply carried away to the sea. In Moscow, the shallow and slow Moskva River could not provide a sufficient reservoir for the offal of meat production. As the members of the Slaughterhouse Commission concluded: "the slaughterhouse brings no harm only if it is kept clean," and "in light of contemporary knowledge, it cannot be allowed to discharge the waste waters from the slaughterhouse straight into the river, without filtration or decontamination."[20] As the project of the municipal sewage system for Moscow was still at the discussion stage, it was proposed to build a separate small sewerage for the slaughterhouse, which would bring its refuse to the filtration fields to be organized at a large wetland southeast of Moscow. The filtration fields of the slaughterhouse, in the opinion of commission members, were thus meant to serve as a kind of testing platform for what would be entirely new in a Russian system of urban waste treatment.[21]

Slaughterhouse Reform in Municipal Discussions

The construction of the central slaughterhouse according to the new scientific imperatives was a complicated and expensive undertaking. The costs of the complex were estimated at 1.9 million roubles—a sum that amounted to almost one-half of all annual municipal revenues in the early 1880s.[22] As one municipal deputy claimed when the project was discussed in the city council:

> In light of the unsanitary conditions in which the urban population lives, the universal pollution of soil and groundwaters, the existing [private] slaughterhouses do not exacerbate the awful unsanitary state of Moscow. Considering the absence of public services in the city, the organization of the new slaughterhouse can be compared to the following: we were given a man, sick from eternal dirt, crippled, in rags, uncombed and hungry and were told to put him in order—but instead of cleaning, dressing and treating him, we would only wash his feet, only toes, and give him shiny shoes. In my opinion, the slaughterhouse is no more than shiny shoes in the matters of urban accomplishment. The slaughterhouse is just a detail and cannot be as important and urgent, as the enterprises necessary for general infrastructure and health of the city, such as water supply and the sewage system.[23]

When so many spheres of urban life required municipal intervention, it was questionable whether the efforts and resources should concentrate on the production of meat. Yet moving the slaughterhouse to the top of the municipal agenda, ahead of the sewage system, had several important advantages. The most obvious was the much lower price of the abattoir construction and the promise that the enterprise would eventually pay off the investment.[24] Another important aspect was that, in the 1880s, there existed a certain consensus in the expert community as to how and why the abattoir should be constructed, while the field of sanitary engineering and waste treatment was painfully adjusting to the new bacteriological discoveries. In this context, the abattoir appeared a more feasible and worthwhile undertaking that would allow municipal leaders not only to quickly accomplish an important sanitary reform for the common good (and take credit for it) but also to gain experience in large infrastructural projects before approaching the much more complicated sewage system.

Although not supportive of the project, the above quotation reveals how important the "common good" rhetoric was for slaughterhouse construction. Similar to other European cities (and different from the American experience, where large meatpacking plants were running for profit and serving markets across the country and beyond), the centralization of slaughtering in Moscow was being justified by public health improvement in a specific urban community.[25] As the members of the Slaughterhouse Commission warned in their project:

> If we admit that the aim is not in material profit but in the desire to protect the city from the harm arising from the uncontrolled meat supply and the upkeep of slaughterhouses in the conditions, incompatible with the elementary notions of cleanliness, as well as to shield the city and its suburbs from the epizootics, we have to agree that this aim can only be achieved at the expense of the material profits of production. Certainly, better veterinary and sanitary control, cleaner upkeep of the slaughterhouse, [and] faster removal of waste mean higher costs and, consequently, lesser crude income of the enterprise.[26]

This argument targeted not only private butchering facilities but also the management of the municipal slaughterhouse by a private concessionaire, because, in the words of Leonid Sumbul, one of the authors of the project, "an entrepreneur is always inclined to gain maximum profit and to avoid the sanitary rules" and "the dirtier the slaughterhouse is, the less expenses it requires."[27] The members of the commission thus concluded, and the majority of the city council agreed with them, that the municipally run slaughterhouse was the only way to reach the public health goal.

Furthermore, the mere assumption that the abattoir could potentially become a profitable enterprise was used as an argument against and not for its construction. A group of deputies opposed turning the production of essential goods into a source of municipal profit and thus placing additional burden on the city dwellers. One such deputy, Nikolay Lanin, himself a factory owner, warned:

> If we see the slaughterhouse as a profitable enterprise, the revenues of which will come to the municipality from the poor consumers, we need to admit that this principle is perverted, that it does not match the status of the city deputies, whose mission is to protect the interests of the majority that they are meant to represent. Therefore, if the main motive for the construction of the slaughterhouse is that it would be a profitable commercial enterprise, I am against this construction. We all complain about the expensiveness of the city life, but it arises from a sum of circumstances that we are in power to change because the wise lawmaker gave us this possibility.[28]

The new public abattoir was imagined not as a correction but as an antipode to the existing businesses in the sphere of meat production. If private facilities were small and dispersed, the new one should be large and centralized. If private enterprises were running for money, the main rationale for the municipal one was common good. Whereas the existing slaughterhouses were dirty, fetid, and full of rotting waste, the abattoir was a clean and hygienically kept place where pure and abundant water carried all the refuse away to filtration fields. Private slaughterhouses endangered city dwellers by letting out contaminated meat, whereas the new abattoir mobilized the achievements of veterinary medicine and sanitary engineering to protect the health of the urban population. The private slaughterhouse was all about disorder; the new abattoir was "rational" and "scientific."

To reach the private slaughterhouses, the cattle were driven through the streets of the city, exposing the population of adjacent neighborhoods to the sight, smells, and sounds of the animals and reminding the public of their inevitable death. In the new public abattoir, as veterinarian Valentin Nagorskii formulated it in his note to the project, "the turnover of animals should be confined to the most limited space, while all the time spans between unloading and arrival at the stockyard, between leaving the stockyard and the slaughter . . . should be cut to a minimum."[29] The new abattoir thus would enclose not only the circulation of animals but also any awareness of their transition from life to death, making that transition invisible and inaudible.

Besides, the public abattoir was seen to be so irresistibly European. Although the circulation of knowledge and practices and borrowing from foreign models was a common feature of the time, the entire discussion of the project in the city council was embedded in the narrative of Moscow's perceived "backwardness" compared to Western cities.[30] The speakers invoked Moscow's "universal pollution," unsanitary conditions, disease, and poverty to call upon council members to "make a step toward the improvement of the city so that it resembled a European one."[31]

Quite illustratively, one municipal deputy mentioned the English law of 1486 against the organization of slaughterhouses in the cities to claim that Moscow was four hundred years behind in resolving the question.[32] Although, in fact, the City of London banned private slaughterhouses only in 1927, and throughout the nineteenth century British butchers successfully opposed the introduction of public abattoirs.[33] Foreign achievements in urban improvement presented a challenge and a motivation for slaughterhouse reform. References to European cities such as London, Paris, Berlin, Vienna, Geneva, and ultimately, Saint Petersburg were thus used to emphasize the necessity and the urgency of the abattoir's construction and its priority over other infrastructural concerns.

To ensure the rationality and the proper scientific basis of the slaughterhouse it was decided to commission three independent projects. The winning design was chosen for its detailed attention to the sanitary side, including waste removal and filtration fields. That project profited from various spheres of expertise, both in Russia and abroad; its authors consulted Russian hygienists, veterinarians, sanitary physicians, and meat producers and made a study trip to visit the public abattoirs in Berlin, Hanover, Brussels,

FIGURE 7.1. Moscow municipal abattoir. From *Al'bom zdanii*. National Electronic Library (Russia).

FIGURE 7.2. Moscow municipal abattoir. Cattle market. From *Al'bom zdanii*. National Electronic Library (Russia).

Paris, and London. In May 1886, the Moscow City Council approved the project and the work of construction began.[34]

Science, Technology, and the Public Image of the Abattoir

The construction process of the new public abattoir was in itself remarkable. The large complex of fifty buildings and complicated infrastructure, most of which was new in Russia, was built in less than two years. The size of this complex was about 30 hectares; the abattoir's sewage farm and filtration fields took up another 150 hectares (see figures 7.1 and 7.2). The efficiency and speed of construction as well as the relatively low financial overrun could speak for the strong commitment of the municipality and the project implementers to the cause of the public good rather than personal material profit.

The abattoir was officially opened in June 1888, yet it did not start operation until mid-August when cattle were finally redirected to the municipal stockyard, giving the abattoir a competitive advantage over private slaughterhouses (which were never officially forbidden but eventually closed in 1892 by an administrative decree in connection to the cholera epidemic). In the first three days of its proper operation, the new Moscow abattoir processed 5,312 head of livestock. "This was how," wrote Dmitrii Gorbunov in his volume for the twenty-fifth anniversary of the Moscow abattoir, "factory production in the sphere of animal slaughter began."[35]

"Factory" was indeed an appropriate word. Everything was done to turn slaughter into an industrial process. It became highly technological and dependent on complicated mechanisms and engineering structures. A separate railway line was built to bring animals to the abattoir. Special transporters, rails, wagons, winches, and lifts moved their bodies and then their carcasses inside the abattoir. The water from a ground pumping system washed away the blood and manure to sewers where a combination of flush tanks and ejectors carried it to the filtration fields. Fans and filters ensured proper ventilation, and steam engines were used in the central heating system and refrigerators. Microscopes helped detect dangerous organs and carcasses, which were then sent to shredders and sterilization machines.[36]

The public abattoir was in many ways a Western product on Russian soil. The idea of it was borrowed from Western Europe and inspired by European examples. Study trips to Germany, France, Belgium, and Britain and consultation with foreign experts facilitated knowledge circulation and direct transfers. The Moscow abattoir absorbed the expertise and experience of several European cities. It was constructed according to the French system, where each function was performed in a separate building.

Its refrigerators were built on the model of Hamburg slaughterhouses. The Delacroix sterilization machines used in Moscow were invented by a veterinarian at the Antwerp abattoir and developed by German engineers. The hydropneumatic sewage system of the Moscow abattoir was first devised by Isaac Shone and successfully used in several British cities.[37]

Although it was mostly the ideas and plans that traveled from abroad, the realization in actuality remained in Russian hands. Local engineers prepared the final project of the slaughterhouse and stockyard, and scientists from the Moscow Agricultural Academy planned and organized the sewage system and the filtration fields. A Russian industrial company Dobrov & Nabholz produced most of the equipment and invented the system of lifts and transporters used within the slaughterhouse, which allegedly made the killing process there faster and easier than in its Western prototypes.[38]

Yet it was not the industrial production of meat but sanitation and health goals that the municipality invoked to create a public abattoir, and the veterinary inspection had to become a crucial part of its operation. The organization of the proper meat quality control was, however, a difficult task. According to the report of the abattoir veterinarians, in the early years "the organs of the killed animals were piled on the floor of the slaughter chamber which immensely complicated their inspection, at the same time allowing the butchers to cut off or hide the damaged parts, and often made it impossible to identify the carcass of the infected organ." In 1891, the introduction of new devices for hanging and numbering the organs and carcasses and the invitation of additional personnel allowed veterinarians to concentrate on inspection and to start individual registration of all the pathologies, regardless of whether they caused the rejection of meat or not.[39]

The inspection of swine and pork was set up to be more effective. The key reason for that was the fear of trichinosis, which was also among the crucial arguments for the centralization of meat production and the ban on private slaughtering. At the Moscow abattoir, from the very beginning, meat samples from every pig were sent to the microscopic laboratory. Although actual cases of trichinosis were extremely rare, this policy favored better detection of other pork parasites.[40] The meticulous inspection of pork was facilitated by its relatively small quantities—it comprised less than 10 percent of all the meat produced in the Moscow abattoir.[41]

The volatility in the numbers of rejected products (see table 7.1) reflects the complicated process of formulating veterinary policy at the Moscow abattoir. The withdrawal from sale of dangerous or unhealthy animal products was seen as a key task of the abattoir's veterinary organization. Yet the category of "unhealthy" remained vague, not only because knowledge

TABLE 7.1. **Cattle Morbidity and the Rejection of Slaughter Products at the Moscow Abattoir.**

	1889	**1890**	**1891**	**1892**	**1893**	**1894**	**1895**
Killed	147,769	153,591	171,142	158,499	161,700	171,829	177,815
Diseased animals (number)	2,647	5,005	7,376	87,190	114,470	119,753	130,174
Diseased animals (proportion)	1.8%	3.3%	4.3%	55%	70.7%	69.7%	73.2%
Morbidity (cases): bovine tuberculosis	776	1,726	2,978	6,759	9,038	12,487	15
actinomycosis	448	596	560	2,419	9,668	8,798	9,640
rinderpest	6	–	1	–	2	–	–
anthrax	–	–	–	3	–	3	1
foot-and-mouth disease	14	79	65	114	351	203	17
Rejected carcasses	421	330	948	1,359	1,279	612	526
Rejected parts	4	-	-	8	100	418	785
Rejected organs	2,152	4,675	6,428	17,798	26,172	29,165	30,321

Source: *Veterinarnyi nadzor.*

of many diseases was limited but also because it was unclear with which health—animal or human—this category operated.

The imperial medical legislation was of little help to Moscow veterinarians as it did not, with some rare exceptions, stipulate any disease-specific measures, stating simply that the meat of sick animals should not be used for food.[42] Rigorous compliance with the law would have meant the rejection of two-thirds of all the abattoir's output because in the veterinary statistics of the 1890s about 70 percent of all slaughtered animals were labeled "diseased." On the one hand, those numbers presented a powerful justification of the necessity of the public abattoir and the veterinary inspection. On the other, they could also reveal the veterinarians' own devotion to their science and their willingness to contribute to advancing fundamental knowledge of animal diseases rather than merely preventing low-quality meat products from entering the market. For veterinarians, the category of "sick animals" was about the health of the animals as such, while the imperial and municipal legislation on the sale of meat used a narrower category, where animal health mattered only to the extent to which it could affect the health of humans. This discrepancy of interpretations created a certain ambiguity in how the "healthiness" of the product was established and how the "unhealthy" products were confiscated.

The confiscation of meat products was generally a new phenomenon in Moscow's slaughtering business, and in most cases it provoked resistance among the livestock owners. As the abattoir veterinarian K. Z. Kleptsov wrote to the municipal Public Health Committee in 1891: "among the external circumstances that until now substantially hinder the correct organization of the sanitary veterinarian business at the abattoir is the complete lack of discipline [*nedistsiplinorovannost'*] among the cattle owners and their managers. Forced to come to the abattoir from the private slaughterhouses, which did not have any sign of control of the product quality, the owners naturally tried to bring their old methods here. Their clashes with the veterinary control proceeded from their being unaccustomed to the sanitary requirements and, on the other hand, from their economic interest."[43]

The abattoir in the early years of its operation did not have a monopoly on slaughtering, so it had to adjust its sanitary goals to the economic interests of its clients. This pressure forced municipality to introduce a generous compensation policy according to which the owners of any confiscated meat received 70 percent of its market price. In fact, members of the municipal committee on the exploitation of the abattoir, seeing the confiscation of unhealthy meat as a part of the sanitary reform, proposed full reimbursement, but the city council rejected the proposal out of fear that this compensation policy would encourage the trade in sick animals and attract too many of them to the city.[44]

At the same time, the necessity to compensate for confiscated products made the municipality and its veterinarians look for ways of reducing the amounts of rejected meat, at least in cases where the animal diseases posed no known risk to human health. In the first years of the abattoir's operation, the detection of disease often meant the destruction of the entire carcass, whereas later on usually only certain organs were withdrawn from sale. The turning point can be identified in the early 1890s, and it is connected to certain changes in the operation of the abattoir.

Although the municipality never received the right to legally impose the centralization of slaughter, in 1892 the new Moscow governor-general, Grand Duke Sergei, ordered the private slaughterhouses to close in the view of their unsanitary condition.[45] This measure was a part of the emergency sanitation campaign by Moscow administration in fear of the raging cholera epidemic in Russia. Despite its temporary character, it allowed the Moscow abattoir to de facto monopolize animal slaughter and to receive an advantage in its interactions with livestock producers. Meanwhile, the abattoir's compensation policy started raising serious concerns—in the first five years the reimbursements consumed about one-ninth of the enterprise's gross revenues. Using a statement from the Medical Society of Moscow University that the localized pathological processes should not involve the rejection of the entire carcass, but only specific organs, and predicting the reduction in the amount of confiscated meat and the financial losses of the livestock owners, the city council eventually canceled all compensations from May 1894 onward.

This change in policy also marked a difference in attitude toward the bodies of animals meant for consumption and those meant for production. Thus, in the same year of 1894, the city council passed regulations on the compulsory killing of animals with glanders, bovine pleuropneumonia, and rabies. Their bodies were seen as valuable to their owners who were entitled to compensation of up to two-thirds of the animal market price. The confiscated bodies of animals that owners already meant for slaughter, on the contrary, were regarded as a mere commodity and were treated in line with other food products. Talking about the bodies of diseased animals at the Moscow abattoir, the opponents of compensation argued that "there is no reason to make exceptions for one type of comestibles and give a compensation that looks like a reward for bringing unsound commodities to Moscow."[46]

The abattoir with its veterinary organization emerged as a kind of shield, protecting the health of the urban population from "uncultured" and "undisciplined" livestock owners. They were seen not as partners in the task of city food supply but, rather, as adversaries whose interest could be neglected, who had to be controlled and converted to the new faith of public health, although its doctrine was not yet established and underwent modification every year. With the rapid development and professionalization of veterinary medicine in Russia, the knowledge of animal diseases, their classification and treatment as well as the perceptions of risks they posed to human health were all subject to constant change. For example, in the 1890s, the international scholarly reevaluation of bovine tuberculosis resulted in a completely different approach toward the treatment of meat from animals with that disease. If in the early years of the abattoir's operation any detection of tubercles led to the rejection of the entire carcass, the new rules of 1895 prescribed that only the organs in which the pathological process was localized were to be destroyed, while the rest of the carcass was sanctioned for sale.[47] The elaboration of more precise norms of meat quality and rules of control allowed for a significant reduction of rejected carcasses. If in the first five years of the abattoir's operation on average 826 carcasses were rejected each year, by the turn of the century this number had dropped to 113, despite the substantial increase in absolute numbers of slaughtered animals.[48]

The control and expertise of the abattoir's veterinary organization, in fact, reached far beyond the city it was meant to serve. The effective inspection of dissected animal bodies in Moscow revealed what local veterinarians in the southern provinces of the Russian Empire had overlooked. In 1893, Moscow veterinarians informed the local authorities of Kharkov about the cases of rinderpest in the herds coming from that province, which helped prevent this dangerous epizootic on the spot. Similarly, frequent detection of tubercular animals at the Moscow abattoir undermined the widespread belief that, unlike in Western European countries, the cattle of the Russian steppes were free from bovine tuberculosis. In the words of its veterinarian, the Moscow abattoir emerged as a "station for the control of the veterinary-sanitary condition of the stock raising in the vast region of Russia that sends its cattle to Moscow."[49]

The presumed scientific role of the abattoir had influenced its construction from the very beginning. Commenting in 1885 on the project, Valentin Nagorskii emphasized the importance of studying animal pathologies at the abattoir: "Livestock, particularly steppe livestock, and its diseases have so rarely become an object of scientific studies—although these studies could give valuable knowledge to science and practice—that it would be very much desirable to organize at the abattoir a laboratory and a museum: the first one to conduct scientific research in the field of animal pathologies, the second to collect and preserve all those rare pathologies with which neither practitioners, nor scientists can work at the moment."[50] The implementation of these recommendations was probably helped by the fact that Nagorskii was personally involved in the project discussions and organization of the abattoir's veterinary control. The abattoir received a laboratory for research and a museum that held "the only in Russia" collection of waxworks for the study of meat, preserved examples of animal pathologies and parasites, exhibits from slaughter-related industries, and statistical materials, maps, diagrams on morbidity, rejection, and so on. Both the laboratory and the museum contributed to the scientific reputation of the institution and served as models for veterinary organizations in other parts of Russia.[51] Furthermore, as veterinarian Nikolai Zelenin wrote in his study of the Moscow abattoir, "the laboratory examination of the slaughter products allowed the veterinary organization of the Moscow abattoir to put the inspection and rejection of meat on a strictly scientific basis. This not only offered better guarantees to consumers regarding the quality of meat on the market but also saved the livestock owners from unnecessary losses because it eliminated rejection on suspicion. In addition, the systematic laboratory research of certain pathologies allowed the Moscow abattoir to become the first in Russia in detecting cases of anthrax and other dangerous diseases that had previously eluded control."[52]

Indeed, the large quantities of empirical data empowered the Moscow abattoir to become a center of research in animal diseases. This was also helped by the newness of the field and the lack of scientific studies, established rules, and elaborate legislation. In some cases, looking for the authoritative opinion in the field, the municipal veterinarians asked the Medical Department of the Ministry of the Interior for instructions on how to deal with cases of disease for which no specific legal regulations existed. Although the responses of the Medical Department—which sometimes arrived months after the initial request—referred to the cases of epizootics in specific animals or herds, the abattoir veterinarians used them as legal precedents to formulate general regulations.[53] Arguably, the sluggishness of the bureaucratic system and the inability to consult with imperial medical bodies on the resolution of every pressing question gave the veterinarians of the Moscow abattoir more agency and power to construct their own norms and to turn their expert knowledge into policy, which was later adopted by public abattoirs in other cities of the empire.[54]

The scientific importance of the abattoir, its complicated technology, and its role as a sanitary enterprise shaped and defined its image in public eyes. Historians have pointed out that in Western cities, particularly in Britain and France, the shift toward new public abattoirs not only made killing invisible and anonymous but also led to the cultural marginalization of the slaughterhouse itself, its exclusion from everyday life and its transformation into a "no place." In this respect, the Moscow abattoir followed a different path.

From the very beginning, it was meant to symbolize municipal commitment to the goals of public health and to be a step on Moscow's way to becoming a "European city." Indeed, seeing the slaughterhouse as a technological and scientific masterpiece, municipal leaders turned it into a center for promoting science and education. Apart from the laboratory and the museum, the slaughterhouse complex had a three-hundred-seat auditorium for scholarly lectures and hosted national exhibitions of cattle raising and butchering.[55]

The Killing Factory

If in France and Britain, the brutality of slaughter was mitigated by using the euphemism *abattoir* (and despite the fact that I have been using this term throughout the text), the original Russian word *boinia* kept the most direct reference to slaughter (only in the Soviet times the institution changed its name to a more neutral "meat complex" and kept it until today). In Moscow, the function of the abattoir was highlighted in several new toponyms that emerged around it: the railway station in its vicinity was named Gorodskie Boini (City Slaughterhouse), then Skotoprogonnaia ploshchad' (Cattle-Driving Square) between the railway platform and the abattoir unambiguously continued with Miasnaia-Bul'varnaia ulitsa (Meat Boulevard), which led downtown.

Russian health reformers, city deputies, medical scientists, and journalists did not display any embarrassment or moral concerns about the presence of slaughter in the city; it was not masked but emphasized. Every city map clearly named the abattoir and many depicted it in detail. The municipal journal each month devoted dozens of pages to its work, and the city guidebooks advertised it as "one of the most remarkable city institutions" and "grandiose construction," "built according to the newest scientific requirements."[56]

Similar to its West European prototypes, the Moscow abattoir was meant to remove blood and death from the city and to confine them within its walls, turning killing into a scientific and strictly controlled process. Avoiding the eyes of the city dwellers, the trains brought the cattle from remote provinces straight to the slaughterhouse where it was let out in the form of meat, lard, leather, or bone meal. The by-products and wastes of that transition were sterilized, recycled, or removed with the help of the complex sewerage system to the filtration field so that the urban public were spared not only the sight but also the smell of slaughtering.

Yet it was exactly the scientific success and the technological innovativeness of the project, especially in a city that was striving to catch up with Western metropolises, that prevented the marginalization of the slaughterhouse. The Moscow abattoir was simply too good to become a "no place." In the eyes of the Russian public, it was an archetype of modernity: conceived by the public self-government body, it consolidated technology and science for the sake of common good, public health, and social progress. And thus, as a successful, profitable, and modern municipal institution, it deserved to be named and its presence within the urban space had to be acknowledged. Instead of turning into a "no place," the Moscow abattoir became a site where the rationalized, mechanized, and sanitized transition from living animal to edible meat was a source of pride rather than discomfort.

CHAPTER 8

Civilized Slaughter

Animal and Human Welfare

The rise of veterinary medicine, sanitary reforms, and the establishment of the public abattoir led to the medicalization, commodification, and industrialization of the human-animal relations in the context of meat production. These processes had, however, another important dimension. The "civilizing mission" of the international abattoir reform was not defined only by sanitation, rationalization, maintenance of public order, and the promotion of scientific and technological knowledge. It also included a humanitarian aspect. In modern societies, empathy toward animals became an important cultural value and a sign of civilized behavior and morale. The recognition of animals' susceptibility to suffering, even if they were denied the ability to reason and talk, drew them closer to people and prompted concerns about human responsibility to prevent their suffering. In fact, animal lack of speech became an important argument of the nineteenth-century animal protection movement as a reason for the human moral duty toward those "who cannot speak for themselves."[1]

This chapter discusses animal and human welfare in the context of slaughter. Fortunately for a researcher, the abattoir, like many other public centralized institutions and unlike private slaughterhouses, left abundant archival documentation, which provides fascinating insight into the history of Russian society and its relations with livestock animals, allowing us to reconstruct practices and attitudes that otherwise would have escaped historical memory. What was happening inside the abattoir? How did the slaughterhouse reform affect relations between humans and animals in the process of killing? How was that industrialized and sanitized slaughter experienced by humans who worked there? And, considering the importance of the "public good" rhetoric for the slaughterhouse reform, how far did the humanitarian goals of the abattoir go and what did they mean for animals and for humans?

Violence or Torture?

In the middle of the nineteenth century, animal protection became a fashionable occupation among social elites and the growing middle classes and went along the lines of other humanitarian or charitable activities such as poor relief or children welfare. The first Society for the Prevention of Cruelty to Animals was created in Britain in 1824, and in the following decades similar organizations appeared across Europe and the United States.[2] Russian elites were no exception in this process. Not surprisingly, animals emerged as a matter of concern in Russian society during the Great Reforms, which stimulated civic activism and put the questions of humanity and legality in the focus of public attention. The first society in Russia, the Society for the Protection of Animals (RSPA, Rossiiskoe Obshchestvo Pokrovitel'stva Zhivotnym) was founded in Saint Petersburg in 1865 and then quickly expanded to other cities. The goal of the society, as formulated in its charter, was to prevent the cruelty and maltreatment of animals—through petitioning the government about appropriate administrative and legislative measures, reporting cases of torture, advocating for the promotion of veterinary medicine, the improvement of slaughterhouses, and the "encouragement of compassion to animals, particularly among the common people."[3]

The main achievement of the RSPA was the criminalization of cruelty to animals in 1871, codified in Article 43 of the Statute of Magistrates as "causing wanton torment to domestic animals," for which the guilty were subject to a fine of up to ten rubles. The success of the society in effecting a legislative change derived from its close affiliation with the Ministry of the Interior and the assistance it received from the central government.[4] The RSPA, in fact, enjoyed remarkable support from the traditional social elites. It existed under the royal patronage of Grand Duke Nikolai Nikolaevich the Elder (brother of Alexander II and uncle of Alexander III) and after his death in 1891 of Grand Duke Dmitrii Konstantinovich (cousin of Alexander III), who in 1902 had to cede this position to the Dowager Empress Maria. The members of the society included senators, governors, representatives of aristocracy, and high church, army, and civil officials.[5]

In 1892, the president of the RSPA, S. Nikiforov, wrote in his editorial to the society's annual report that cruelty to animals can serve "as a correct indicator of the coarseness and ignorance of a given people" and cannot be found "among a developed and civilized people."[6] The aspirations of civilization demanded banning cruelty to animals. The charter of the RSPA explained cruelty and maltreatment of animals as overworking, failure to provide food and shelter, torturing them on a whim, and "cruel treatment of animals during slaughter."[7] But what exactly was seen as cruel in an institution whose entire purpose was mass killing?

In 1896, members of the RSPA became interested in the activities of the Moscow abattoir and revealed numerous cases of what they described as "animal torture," initiating a discussion on animal welfare with the Moscow city government and the abattoir management. The protocols of the RSPA, its petition to the city mayor, and even more so the responses of the abattoir officials, all present unique sources that help us reconstruct the various perceptions of slaughter held by insiders and outsiders. They reveal human attitudes toward animals and, eventually, toward humans themselves.

Internal rules of the Moscow abattoir regulated the treatment of livestock and allowed for two forms of violence—beating animals with a stick or twisting their tail to make them move into the slaughterhouse. However, the protocols of the RSPA reveal that those norms had little to do with reality and that animal abuse was a part of the abattoir's daily routine. The livestock, as the protocols of the RSPA described, were constantly beaten and often left without food, water, and shelter for days. The stockyard was paved with sharp stones that injured animals and prevented them from lying down. The piglets and calves were jam-packed when transported; they were tied too firmly and generally treated "like bags."[8]

Interestingly, the RSPA did not disapprove of the act of animal slaughter per se—although voices against animal consumption were gaining strength in Russian society. The most famous call was Leo Tolstoy's 1891 article "The First Step," which argued that the rejection of meat consumption should be the first step toward any moral and spiritual development of mankind.[9] The right of humans to kill animals on an industrial scale was taken for granted by the RSPA, it was only the form that mattered. In fact, in certain cases (with injured or old animals) killing was seen as a humanitarian deed, a kind of coup de grâce.[10] Only the emotionality of the description reveals that the RSPA investigators must have found the sight of slaughter quite shocking: "The bull jibs, refuses to go in the doors of the dark slaughterhouse with the vapors from the streams of fresh blood coming from there; the striker squeezes and breaks the tail of the bull and severely hits its sacrum, sides and legs with a heavy cudgel.... The human heart fills with compassion when one remembers the poor animal which is simultaneously strong and helpless, seized with horror and tortured with merciless hits."[11]

To clarify the situation, the RSPA invited to one of its meetings the chief abattoir veterinarian Gavriil Gurin who confirmed that society members' observations were correct and mentioned even more cruel examples such as cutting hindleg tendons to prevent bulls from jibbing ("to wear bast shoes" in slaughterhouse argot), pulling the skin off the bull's tail with two sticks ("fiddle"), or putting a stick inside the bull's anus to make it move faster. The fact that those practices had specific names in the professional language of workers suggests that these were not exceptions but common methods of preparing for slaughter.[12]

Despite the stunning examples of animal abuse, documented and even photographed by members of the RSPA, their proposals, formulated in the petition to the Moscow mayor, were rather moderate. They called for the better organization of the food and water supply, some logistical improvements in the transportation of livestock to and inside the abattoir, repaving of the stockyard and provisions for the prompt killing of injured animals. An important part of their proposal was also aimed at sparing the livestock from knowing their destiny: the RSPA suggested that animals should be kept in pens with high walls and moved by special workers whose clothes and hands had no signs of blood.[13] This interpretation of animal awareness of coming death as a type of torture deserves closer attention, and I will return to it later in this chapter.

Considering the overall influence of the RSPA, its petition could not remain without a response. Moscow city board demanded an explanation from the abattoir management, and the Central State Archive of the City of Moscow preserved two responses. One was from D. Verderevskii, who worked as the abattoir's managing director since its creation; the other was from the senior veterinarian Gavriil Gurin on behalf of the abattoir's veterinary organization.[14]

The response by Verderevskii was remarkable not only in its length and eloquence but also in the surprising level of reflection about the purpose of the abattoir and the human-animal relations within it. His response is worth discussing here in detail. As the RSPA framed its petition around the claims of animal torture, Verderevskii started his response with questioning the entire concept of torture:

A man subdues an animal, not gifted with reason, only through violence [*nasilie*]. The animal does not voluntarily give its body to the man, so he can only achieve his wish of using the body of the animal through violence, and the most brutal violence—through taking its life, through killing. Should this violence be counted as torture [*istiazanie*]? Obviously not, because this violence is needed to satisfy man's wish to use the necessary animal meat. But to kill an animal, which does not voluntarily allow it, one needs to put it in a condition convenient for the slaughter, to deprive the animal of the ability to resist, and for this one needs to commit violence, more or less cruel, depending on the resistance. Will this violence be torture? Clearly not, because this violence does not proceed from human evil will but from necessity. This violence is necessary because if the man abstains from this violence, he will have to subordinate his rational will to the animal will. Obviously, there are many types of human violence toward animals, often very cruel, which, despite its cruelty, cannot be considered torture because of their utility to men and the absence of human evil will.[15]

For Verderevskii, the ability to reason was the borderline between people and animals and also the source of unquestionable human privilege and power. Even more so, exercising violence over animals was presented as a means of defending the superiority of rational mankind, and the abstention from violence was interpreted not as a victory of human empathy but as a defeat of human reason and the acknowledgment of weakness. In this rationalist explanation, its utilitarian character of violence served both to justify and to encourage it—as one of the ways of bringing nature to the service of people.

Its usefulness distinguished good violence from bad, the rational from the irrational, or rather, violence from torture. From this point of view, what mattered was not whether something was cruel or not but whether it was necessary to humans or not. According to Verderevskii, even the most basic needs of animals such as food were to be satisfied only to the extent that those needs were useful to rational humans. Responding to criticism from the RSPA investigating committee, Verderevskii wrote, "The committee would have been right if men reared livestock only to please it with feeding but since livestock is raised for other purposes, it is fed only to the extent it is necessary for the goals of its owners. Non-working livestock is fed only to the extent it is necessary to keep it alive and partially healthy, and not to the extent of its appetite."[16]

The ability or inability to reason turned particularly important in the question about animal awareness of the coming slaughter. The RSPA interpretation of the animal anticipation of killing as torture, which projected on animals the human fear of death and the capacity for imagination, was of course very anthropomorphic. Through likening human and animal fear of approaching death, the RSPA proposal becomes strikingly emphatic. Verderevskii replied:

> The arguments presented by the committee would be undoubted if the bull was gifted with reason. If we show the bull the cut bull's leg, will it understand what it is? Can it form an idea of the whole from a part of the whole? Clearly, not—for this one needs to have reason, which the bull does not have, therefore, when it sees the organs of the killed livestock, blood, skins and so on, it, unable to reason, can't connect in its mind the impressions from these objects to the image of the living bull and, therefore, to understand that his comrade was killed and lacerated here.[17]

Even today there is no definite answer to the question as to whether animals can feel the coming death; it is still impossible to fully retrieve their experiences at the slaughter. Some recent studies suggest that the cattle resist not because they sense the approaching death—which indeed requires certain engagement with the abstract—but because they are upset and scared by specific disruptions of their visual, olfactory, and auditory worlds that are almost unavoidable within the slaughtering facilities. Taking animal perception seriously can make their experience of slaughter less frightening and thus the entire process more humane—at least as much as its purposes allow.[18]

Verderevskii, however, used the animals' inability to generalize as a pretext to completely dismiss animal experiences and thus remove any grounds for compassion. He connected animal agency and resistance to their irrational, lazy, and stubborn beastly nature, particularly in the case of animals that were never used for work within human households. "This animal does not know what rein is, it equally resists wherever man takes it, and with these animals one is forced to exercise violence when bringing them inside the slaughterhouse—it once again proves that not the circumstances of the slaughter provoke the resistance but only their obstinate, untamed will. . . . Tamed animals go to the slaughter without any fear and resistance."[19]

In his interpretation, force, beating, and squeezing tails, techniques used to bring animals to the slaughter, become necessary and useful as they facilitate the animal moving within the abattoir. In the case when a bull stops and balks, Verderevskii writes, a worker "needs to force the bull ahead and thus prevent injuries among the livestock behind it whose movement it hampers; what can happen in this situation even among people who are able to reason is shown by the catastrophe of Khodynka."[20]

From this point of view, cruelty that is considered necessary, that is rationally motivated and produces a useful result, is not torture. For Verderevskii, torture was unnecessary violence, without any utilitarian purpose (here he implicitly refers to the above mentioned Article 43 with its "wanton torture"). Torture is an unreasonable action and thus unworthy of a rational and civilized man. He acknowledged that animal torture had been seen at the abattoir in the former years, but he connected this to the initial lack of education and civilization among the abattoir's workers:

> At first, it was very difficult to harness the workers, all of whom came to the municipal abattoir from private slaughterhouses. At private slaughterhouses, from the early years they are used to dirt, slovenliness, and cruel violence toward the slaughtered livestock and, despite their sincere willingness, could not understand the imposed requirements of sanitation, cleanliness, and stopping the unnecessary violence over animals. Trying to avoid the discontent of the workers and prevent the strikes that were several times prepared among them, instigated by the old banned slaughterhouses, we were forced to shut our eyes to many actions of the workers that did not conform to the civilized abattoir. Much time and effort was spent to reach the current situation with the slaughter; it was necessary to reeducate the workers, to train them to orderliness, sanitation, cleanliness, to stop the unnecessary violence against animals, and to do many other things.[21]

Torture and excessive violence are thus presented as a product of undisciplined behavior—human or animal—that would be absent in a model where animals were tamed and people were civilized. Verderevskii also emphasized that his institution was well organized and well supervised; whatever disorder existed there did not reveal his unawareness or the limits of his control but was, in fact, his conscious concession to the workers. He portrayed the abattoir as modern, scientific, and rational but also as an establishment that turns its employees into more civilized and humane people and enhances the position of human culture versus nature.

The abattoir veterinarian Gavriil Gurin, the author of another response to the RSPA criticism, had, however, a very different opinion of the entire situation. Although he generally agreed with Verderevskii that the municipal abattoir "because of its public character" encouraged better treatment of animals compared to that at private slaughterhouses, he challenged Verderevskii's point about the efficiency of its organization and control as well as its potential to civilize people through the establishment of rational rules. "It would be too much to claim that, in such a short period of time since the slaughterhouse reform in Moscow, the workers and butchers managed to cardinally reeducate themselves, soften their morale, and abandon their long habits of the cruel treatment of animals. Hardly being concerned about that, the workers, especially when workload is pressing, or when the slaughter continues into the evening, when they are sure that nobody can see them, still allow the undesirable treatment of animals."[22]

Gurin, now in an official form, admitted that the torture reported by the RSPA indeed did occur at the Moscow abattoir, and he provided some more blatant examples of animals abuse (such as pulling animal by the eyelid) as well as excessive beating, which, he claimed, was used habitually, even on obedient and nonresisting bulls.[23] However, unlike Verderevskii, Gurin did not think all these notorious cruelties proceeded only from the ignorance and lack of civilization and discipline among the workers who were not able to absorb the rational and humane rules of the modern abattoir. According to him, this behavior was stimulated by the entire organization of the abattoir, with its long working hours and huge workload, when workers needed to kill nine or ten bulls per hour.

Although Gurin and the abattoir's veterinary organization generally agreed with the anti-torture measures put forward by the RSPA and supported their concern for animal welfare, for them it was human welfare that came first: "In the opinion of the Veterinary Organization, all those cruelties in most cases are not the result of the evil will of the workers but are caused and supported by their conditions of work. When these conditions are changed for the better, the cruel treatment of animals at the Moscow abattoir will stop and remain only in the memories."[24] In their own proposal to the municipality, the first point to promote the better treatment of animals was the better treatment of humans. They insisted that the abattoir should be not only a place of humane slaughter but also of humane employment, which was not the case.

The Limits of "Public Good" and the Abattoir as Employer

In the three decades of its operation, the Moscow municipal abattoir employed simultaneously between 300 and 600 people. In 1910, its permanent staff counted about 450 people and 80 more were recruited for the high season.[25] The management of the abattoir was divided into three branches—(1) economic, (2) technical, and (3) veterinarian. These branches were headed respectively by the managing director, the senior engineer, and the senior veterinarian, who were of equal standing and who reported directly to the Moscow Municipal Board. Their tasks were independent from and sometimes even competing with each other, and as the documents of Gurin and Verderevskii demonstrate, they could have different opinions as to the abattoir's organization. The engineers provided the maintenance of the abattoir's infrastructure, the managing director controlled the slaughtering

TABLE 8.1. **Income of the Abattoir Personnel in 1890.**

Position	Monthly Income (rubles)	Housing
Managing director	200	Free apartment
Senior engineer	200	Free apartment
Senior veterinarian	125	Free apartment
Deputy managing director	100	Free apartment
Deputy engineer	100	Free apartment
Veterinarians	100	Free apartment or extra 25 rubles of compensation
Office clerks	50	Free apartment
Workers (depending on profession)	25–40	Free bed in the dormitory
Microscopists (women)	30	No housing or compensation provided

Source: Gorbunov, *Moskovskiie gorodskiie boini*.

process and meat production, while the veterinary organization was responsible for the safety of that meat and compliance with sanitary rules (for their remuneration, see table 8.1).[26]

The workload at the abattoir was heavy because of the volatility of output throughout the year and because it was understaffed. Even the professionals had to work under pressure and in the "atmosphere of constant rush."[27] In the 1890s, the veterinary inspection of the Moscow abattoir involved a maximum staff of 20. Each abattoir veterinarian had to inspect the products of at least 160 slaughtered animals per shift, and the inspection depended not on the capacities of the veterinarians but on the intensity of the slaughter.[28]

The pressure was particularly high on the workers, and even the administrators admitted that workers were overloaded, especially on the days of mass cattle arrivals when the slaughter lasted until late evening. Normally, blue-collar employees at the abattoir had a six-day working week and a working day of between ten and twelve hours, depending on profession and season. The jobs that required round-the-clock maintenance of machines had twelve-hour day and night shifts with four days off per month. The slaughterers started their day at 7:30 a.m. and worked until 9 p.m. with two breaks for lunch (ninety minutes) and tea (thirty minutes). They were paid for eleven months per year; in winter, when cattle arrivals reduced, workers were forced to take one month of unpaid leave.[29]

Overtime work was frequent at the Moscow abattoir, and managers were more eager to compensate it with money than with additional days off. "The consequence of such labor intensity," wrote Gorbunov, "was the utmost exhaustion of workers, which badly affected the quality of work and even sometimes caused refusals to work overtime." In some technical jobs, workers had no right to refuse overtime work.[30]

Yet, despite the physically and psychologically demanding labor and long working hours, the labor force of the abattoir was remarkably stable and loyal to the employer. Only 10 of the 138 workers employed at the slaughterhouse for cattle in 1906 worked there for less than five years. Half of all slaughterers worked at the Moscow abattoir for more than fifteen years, and 51 for more than eighteen years—that is, from the very first months of its operation.[31] One explanation of this loyalty might be the fact that the Moscow abattoir held a monopoly on its business and basically was the only employment opportunity in the city for those whose profession was connected with animal slaughter. There was higher personnel volatility in professions that were not directly related to slaughter and that had a more diversified demand.[32] The other reason was that the Moscow abattoir with its long and exhausting working day was not very different from other employers on the city job market. Finally, free housing was an important factor that tied the workers to the abattoir.

Unlike its Western models but very much in line with Russian factory traditions, the Moscow abattoir provided accommodation for its personnel. Employer-provided accommodation was a typical feature of Russian factories, both in the countryside and in the big cities. Housing at the industrial site was common not only for the workers but also for the white-collar employees and even the factory owners themselves, who often chose to stay next to their enterprises instead of relocating to quieter and greener areas.[33] An important difference is that the abattoir was not a private but a public institution organized by the government, which claimed to be acting on behalf and for the benefit of the city residents—a position that could have affected housing policy as well.

Constructing accommodation for the personnel was planned from the very beginning. Accommodation was devised and implemented as an integral part of the abattoir's project, from the institutional, architectural, and infrastructural points of view. Taking into account the abattoir's location in the city outskirts, the poor level of public transport, and the general housing shortage in Moscow, free accommodation near the workplace was an offer difficult to decline. Entitlement to housing was not equal or based on need; it depended strictly on the position one occupied in the abattoir's employment pyramid. Although the income difference between the upper and lower personnel of the abattoir was quite moderate, the spatial arrangements reinforced social inequalities and clearly delineated the middle classes from the workers.

The site of the abattoir was a rectangle stretching north to south from the city to the railway platforms. Buildings for the abattoir's administration and white-collar employees were located on the northern edge of the complex, the most remote from the slaughtering facilities. They provided spacious apartments with central heating and running water, oak parquetry, and private kitchens and bathrooms. Behind them stood the buildings for the mid-rank personnel such as foremen and guards, with small and more simple apartments and shared kitchens and toilets. The so-called family barrack, located between the margarine and the albumin factories, provided twenty rooms for highly qualified workers (mechanics). These twenty families of the working-class elite had to share two kitchens and two toilets. All other workers lived in the barracks in the center of the abattoir's complex, right next to the slaughterhouses for pigs and calves. Compared to the average dwellings of the Moscow working classes, those barracks looked quite good—solid brick buildings, freshly painted, warm, dry, and well lit, connected to the sewage system and the water pipe, with big windows, heated toilets, and even a marble staircase. What a worker could receive in those modern dormitories of the abattoir was, however, only a place in a room for twenty people where one got a bed, a mattress, a blanket, and a closet for belongings.[34]

The geometry of the abattoir's housing arrangements was a spatial enactment of hierarchies within the abattoir's population. It remained a male world, and women, at least those who were not family members of the higher ranks, were practically excluded. Although two professions—laundresses and microscopists (*mikroskopistki*)—were reserved for women only, they received the lowest salary in their employment groups and were the only personnel for whom no housing (and no compensation) was provided.[35] The cleanest, best-looking, and the most representative edge of the abattoir, with trees and flowerbeds, accommodated the elites responsible for modern science, technology, and efficient economic production. They could enjoy comfort and privacy, at least to the extent allowed by the abattoir's circumstances. The workers, associated with crudeness and violence, lack of reason and civilization, were placed closer to the animals, and their life was subject to constant regulation and control.

The abattoir was a disciplining project in many respects. It was conceived to restrict and impose scientific rules on the "undisciplined" livestock raisers to prevent them from supplying improper meat. It was also meant to discipline the meat consumers, to make them learn what kind of meat was good for them and to abandon unhealthy practices. Similarly, it was also a disciplining project for its workers, who spent their entire days within the abattoir's walls and, unlike the other groups, were supposed to be under constant supervision, either at work or in a dormitory room for 20, or in a canteen for 150. The workers were discouraged from leaving the walls of the abattoir as not only housing but even some forms of leisure were provided inside.[36]

The spatial arrangements of the abattoir exposed not only social differences in the right to comfort but also differences in the right to privacy, domesticity, and family life as such. The possibility of living together with a wife and children was given only to the middle classes and a small group of working-class elite and denied to ordinary workers, whose living arrangement resembled those of soldiers or prisoners. Although workers were not expected to abstain from sexual life, it was channeled into a spatially and temporarily restricted form—the workers' dormitories (housing three hundred men) included four rooms for "brief visits of workers' wives," whose stays could not last longer than ten days.[37] Remarkably, this was not an opportunistic or accidental decision on the use of empty premises nor a temporary solution to the housing problem but a preconceived, well-thought-out project, approved by the municipality, which suggests that this type of working, housing, and family organization was perceived as appropriate for workers and did not seem to contradict the municipality's vision of public good.

It was not until the revolutionary year of 1905 that working conditions at the abattoir improved. Many Moscow municipal delegates generally shared the revolutionary demand for large-scale reforms, especially in the political sphere. On January 14, 1905, five days after Bloody Sunday and the beginning of mass protests and strikes, the city council stated that the workers should have all legal ways to protect their interests and spoke in favor of the freedom of strikes, unions, and meetings. In June 1905, the city council adopted proposals on the improvement of the state order in Russia, which called for the creation of the parliament on the basis of the universal, equal, secret vote.[38]

Yet in the sphere that directly concerned the city budget, municipal leaders were more reluctant to make actual concessions to the growing workers' movement. The abattoir's workers were in fact the first among the large group of municipal employees to protest and fight for their rights. In early April 1905, they petitioned the municipality for twelve-month employment with two weeks of paid vacation, for the reorganization of kitchens, more rooms for family visits, and for a general increase in wages—which had remained the same for seventeen years from the very opening of the abattoir. The municipality agreed with most of the demands and ordered management to buy new stoves and to create twenty rooms for family visits; yet the request for a wage increase was found unreasonable. Instead, it was proposed to provide the workers with clothes and boots at city expense. However, in 1905, the strategic initiative was not on the side of the municipality, and boots alone could not satisfy revolutionary demands. It is possible that the abattoir's workers also appeared more threatening to the city authorities than employees of other municipal enterprises—after all, due to the nature of their daily work, these people were not afraid of spilling blood and were skilled at killing. In May, the abattoir workers insisted on a 20 percent increase in wages, and after some negotiation, both sides agreed on a 15 percent increase. In summer, the municipality approved the increase of staff and the two-week paid vacation for all workers and introduced three eight-hour shifts instead of two twelve-hour shifts for those employed in the abattoir's technical maintenance.[39]

Meanwhile, all Moscow workers employed at the various municipal enterprises, including the abattoir, formed a joint workers' organization. This organization, threatening the city with a general strike of municipal employees, demanded minimum monthly wages of twenty-five rubles, awards for long service, limits for overtime work, an eight-hour working day, and one month of paid vacation for all, pensions and insurance against death and disability, improvement of housing conditions, and the right to use their after-work hours at their own discretion.[40] The municipality, facing the general strike and bound by its declared support to the revolution and the workers' movement, had to yield. In October, the Moscow City Council introduced long-service awards of up to 40 percent for all municipal employees, which meant a substantial increase for many of the abattoir's workers. The city government stated that the needs of workers with families should be considered and offered a compensation ("apartment money") for those who might choose to live in private apartments and not in the dormitories. Additional personnel were hired to decrease the workload at the abattoir and the compensation for overtime work increased from 100 percent to 150 percent of the wage. Workers were granted the right to decide personally when they wanted to take vacations, and the maximum duration of wife visits to the abattoir increased from ten to fourteen days per year.[41]

The concession to the workers also meant that council members admitted that conditions at the municipal enterprises had not been good enough, despite all the "public good" rhetoric. As an author of an article in the municipal periodical diplomatically put it in November 1905:

> It has been frequently said in the city council that the municipal government should put its workers in the best conditions because, unlike private entrepreneurs, its goal is not profit but the improvement of life for the city population, to which the workers belong, and because the municipality should give an example of particular care for its workers. Although this thought was implemented to some extent, it is necessary to admit that it was not done systematically, partially because the municipal enterprises . . . were at first unprofitable to the city, partially because of the narrow circumstances with city finances.[42]

Despite being a public project, the Moscow abattoir in its role as employer acted similarly to the private industrial plants. Municipal leaders wanted to distinguish themselves from private entrepreneurs, condemning their focus on profit, and claiming to prioritize public good over economic success, yet in practice financial concerns prevailed. Although city deputies shared the discourse of service to the people and claimed to act for the benefit of the urban community, they failed to see employees of municipal enterprises as part of that community.

The slaughterhouse reform and the appearance of centralized public abattoirs, first in Saint Petersburg and Moscow and then in other cities of the empire, meant that educated elites for the first time became involved in the process of meat production. The circle and the status of those who were familiar with the slaughter changed, and so did discussions about it. From the business of butchers, slaughter became a matter of concern for veterinarians, physicians, engineers, economists, municipal deputies, and aristocratic animal welfare activists. The increased distancing of the public from meat production, which was now excluded from daily urban experience, allowed people to look at the entire process of slaughter with a fresh eye and to formulate a demand for a more humane treatment of animals even when performing the cruelest tasks. This demand and attention to the animal experience during slaughter was new, but the reality clearly remained much more brutal and abusive than the humane ideals of quick and painless death.

One explanation for this was the connection between animal and human welfare. Despite the importance of the humanitarian and "public good" rhetoric in the creation of the abattoir, its policy as employer revealed how narrow the social implications of that "public good" in fact were. The Moscow abattoir was not only a space of mass animal slaughter but also, for hundreds of its employees, a home and a space of work and leisure. As such, the abattoir represented the microcosm of Russian society, exposing its many inequalities and power relations. The life and work of humans there were subordinated to the sanitary and economic goals of the abattoir, whose managers, contrary to early visions from the 1880s, were interested in maximizing output and cutting costs just like private entrepreneurs. Although living and working conditions at the abattoir had been sanitized with new technologies such as the sewage system, toilets, and running water, its workers had for a long time been denied not only decent wages and protection from exploitation but even the right to privacy and family life. The operation of the abattoir shows that public government could hardly offer an alternative to the existing labor and employment relations of Russian capitalism and, in a sense, confirmed that it was only the revolution that could prompt changes in labor policies and improve the living standard of workers.

PART IV

A Paradox of the Sanitary Project

Children in Moscow

CHAPTER 9

A Deadly City for Children

In 1886, the Moscow hygienist Sergei Bubnov—an assistant of Friedrich Erismann and his future successor as the professor of hygiene at Moscow University and the director of the Moscow sanitary station—left the following evaluation of child mortality in Russia:

> The mortality of children under the age of five comprises 59 percent of all deaths and in some provinces up to 76 percent, so this mortality is horrifying. . . . Masses of children are born, masses of children get sick and masses of children die before their time. The population tends to newborns, infants, and sick children and finally buries them in the ground. This enormous labor and enormous expense have never been and cannot be counted and evaluated. Two million and eight hundred thousand persons die in Russia each year, and it is hardly possible to doubt that at least one million of these deaths did not have to happen had there been a proper sanitary improvement, that one million deaths can be avoided and prevented by national sanitary measures, and that every year one million people are buried in the ground for no reason but simply because of some habits and misconceptions.[1]

In the last third of the nineteenth century, Russia, similar to other societies worldwide, witnessed an explosion of interest in children who became an object of scientific research, evaluation, and categorization and a highly politicized matter of national interest. The expert discourse on children, their living conditions, nurture, and education, developed across a number of professional fields, which were also emerging and struggling to define themselves: pedagogy, hygiene, public health, psychology, and psychiatry. New experts criticized traditional practices of child rearing in Russia and tried to formulate and propagate the "proper," "rational," and "scientific" ways of caring for children.[2]

The late imperial decades were also the time of the first efforts to ensure minimal child welfare and to legally protect them from exploitation and abuse. The factory law of 1882 forbade the employment of children under the age of twelve, limited the working day for those under fifteen, and obliged industrialists to provide schools for their child workers. By the turn of the century, child protection within their families also came to the attention of lawyers. The laws of 1891 and 1902 improved the legal status of children born outside of wedlock. The new Criminal Code of 1903 prescribed arrest or removal of parental power for cruel treatment of children under seventeen and for forcing them into beggary, prostitution, or marriage.[3]

An important focus of this increasingly vocal child welfare movement was the extremely high child and (especially) infant mortality in Russia. From the 1880s onward, tremendous loss of life among Russia's children attracted the attention of medical professionals and zemstvo activists, local community physicians, and central medical authorities in Saint Petersburg. Not only were infant mortality rates much higher than in Western European countries with which Russia was competing, but they were also higher than in Central and Southeastern Europe and within the empire itself, most dramatic among the Orthodox population of Central Russia. These mortality rates served as a powerful confirmation that Russian infant and childcare needed radical intervention and reform.

Although contemporaries focused primarily on extreme child mortality in the countryside, death rates among urban children were enormous and, in some age groups, even higher than in the village. Among the metropolises of the Russian Empire and among major European cities, Moscow had the highest child mortality—a poor recommendation for the city that was implementing ambitious sanitary reforms. This part of the book examines the place of children within the Moscow municipal sanitary project and the broader attempt to make the city healthier.

The problem of imperial Russia's exceptional child mortality has been the subject of several historical works, but it has not been examined in the context of broader urban public health programs.[4] The focus of study so far has been either on child deaths in the village (about which zemstvo physicians left abundant and eloquent historical accounts), on industrial provisions for working mothers, or on specific institutions, public organizations, and initiatives in cities that, though important to understand the intellectual foundations and the possible directions of the child welfare movement, usually remained limited in practical influence and, with the exception of the Foundling Homes, affected only a small number of children. I believe, however, that scale and numbers matter and that to understand the experience of late imperial urban childhood (and urban public health reforms), we need to consider the impact that new initiatives in child welfare and health were having on the city population and try to include in the picture the lives and deaths of the many who could not partake in those indeed sometimes innovative but limited projects. Why were so many children dying in Moscow? What were the causes and the dynamics of this mortality? What did the city do or not do about it? What role did the measures aimed at children play in the sanitary project?

Why Were So Many Children Dying?

Child mortality in late imperial Russia was huge by all accounts. Although reliable statistics are absent for the mid-nineteenth century, some researchers note that death rates among children actually rose in the early postreform period and most observers agreed that they were slow to decline. At the turn of the twentieth century, out of 100 children born, 26 died during their first year in European Russia, compared to 22 in Hungary, 21 in Austria, 20 in Germany, 17 in Italy, 14 in France, and 8 in Norway. In the provinces of Moscow, Tula, Saratov and Penza, more than half of children did not live until the age of five.[5]

Medical professionals attributed the problem to inadequate nutrition and care for the children, the necessity for mothers to return to work immediately after birth, and poor provisions for breastfeeding, the too early introduction of solid foods, ignorance of hygiene and the diet of children, overcrowding and poor housing conditions, lack of childcare facilities and medical assistance. Although the experts' evaluations reflected physicians' own struggle for authority, their vision of medicalization as necessarily progressive, and their view of traditional childcare practices as backward, huge infant mortality in Russia was a real problem. The main cause of death among Russian babies, physicians maintained, was gastroenteritis or infant diarrhea, most likely resulting from inadequate food and microbial contamination, which could be prevented by breastfeeding and proper diet, hygiene, and care for the child. The emphasis on nutrition explains why infant mortality was particularly severe among the Orthodox population. According to late imperial observers, death rates among Muslim, Jewish, and

Catholic children of the empire were considerably lower because cultural norms in those communities prevented the separation of mother and child and encouraged and allowed for longer exclusive breastfeeding. In Russian Orthodox communities, on the contrary, women had to return to work very soon after childbirth, especially during the summer months when agricultural work required all available hands. Infants were then left in the care of their older siblings or grandparents, unfit for physical work and often too young or too weak to care for a newborn, and without appropriate food. As a result, infants were fed either cow's milk or solid foods such as bread, potatoes, or porridge in the form of *soska*, when partially chewed food was wrapped in a cloth and put into the baby's mouth.[6]

In the late nineteenth century, most medical attention in Russia concentrated on infant mortality in the countryside where the majority of child deaths occurred. Community physicians saw excessive child mortality as an issue inseparable from the social and economic context of the Russian village, with its poverty, exhausting agricultural labor, and lack of education. Proposed solutions included hygiene information, better access to health care, and the widest organization of nurseries to look after the children while their mothers went to work. In the words of one of the founders of Russian community medicine, Ivan Molleson, the organization of nurseries was "the only measure that can reduce child mortality to its possible minimum" and "has such absolute importance that one cannot go around it."[7]

However, child and infant mortality was a problem not only in the village. In fact, during the first year of life, the situation was the worst in big cities. In 1893–1896, infant mortality in Russian cities was 278 per 1,000 born in big cities, 246 in smaller cities and towns, and 268 in the countryside of European Russia.[8] In Moscow, survival prospects of infants were low even in comparison with other big cities of the Russian Empire, and in the early 1890s more than 30 percent of children died there before reaching the age of one.

What can explain such extreme mortality? One reason was the peculiarity of Moscow urbanization and the significant gender differences of the migrant experience in Moscow, which discouraged the formation of families and favored single motherhood and child abandonment. Moscow's urbanization was characterized by the transient character of migration, which remained predominantly male and featured strong ties of the migrants to their village and the frequent combination of agricultural labor and urban wages within the same family, as well as a high proportion of immigrants and a remarkably low ratio of women in Moscow's population compared to other European metropolises. Although, at the turn of the century, the proportion of women in Moscow was slowly increasing, historians have argued that most of the peasant women who migrated to the city came there for wages rather than in pursuit of family reunification. Migration was often a survival strategy for those women who were marginalized in their communities—orphans, spinsters, widows, or soldiers' wives. They left for the city looking for employment and, perhaps, more freedom and, similarly to their male fellows, sent a large portion of their wages to the village.[9]

The prospects of marriage in Moscow were slim, and marriage rates were significantly lower than in other big European cities. Although the number of men, especially in the twenty-to-thirty age group, by far exceeded that of women, very few of those men were available for marriage, as most had families in the village.[10] Not only were marriage rates low in Moscow, but marriage did not necessarily mean that the woman stayed home and cared for the family. Many married Moscow women continued to work because relying on a single breadwinner and maintaining dependents within a migrant working-class family in the city was extremely difficult, especially when combined with the need to send money back to the village. In 1902, almost 50 percent of female workers at Moscow factories and about 30 percent of female domestic servants were legally married. Overall, 65 percent of immigrant women in Moscow were self-supporting.[11]

As a result of this peculiar social organization and family strategies, at the end of the nineteenth century Moscow had relatively few children. Unlike the village, the urban environment discouraged childbirth and rates in Moscow remained low. However, in Moscow rates of illegitimate children were astonishing—according to municipal statistics, in 1890 about 40 percent of all infants in Moscow were listed as born outside of wedlock.[12] In many parts of the city, a woman was much more likely to become a single mother than to get married; the annual quantity of illegitimate births outnumbered that of weddings.[13]

Employment and living conditions of working-class women in Moscow made it extremely difficult to raise children in the city. Not only was the help of larger kinship networks unavailable to them but also often the resources of an individual family, especially with an illegitimate child. Even if the newborn was lucky enough to stay with the mother, in order to provide for the family she had to return to work as soon as possible, leaving the child without proper food or care. Staying with the mother, however, was not a likely prospect for an illegitimate child at the end of the nineteenth century. Child abandonment in Moscow reached startling proportions. In the 1880s, one-quarter of all children born in Moscow were abandoned in the first weeks of their life and were taken up by the Moscow Imperial Foundling Home (Imperatorskii Moskovskii Vospitatel'nyi Dom).[14] Most of these children were illegitimate. The foundling home accepted only infants, and all those who were orphaned or abandoned at an older age were not included in these statistics.

The existence of the Imperial Foundling Home in Moscow had a huge impact on the city mortality statistics. Created in the eighteenth century to prevent infanticide, the Moscow Imperial Foundling Home experienced an unmanageable increase of abandoned children in the post-reform decades. David Ransel estimates that this institution had to handle a larger amount of abandoned children than social services in any other European metropolis.[15] Although Saint Petersburg also had a foundling home, it remained smaller in scale. In the late 1880s, the peak years of its operation, the Moscow foundling home was receiving annually more than 16,000 unwanted infants—when the number of registered births in Moscow was about 28,000. A large number of these children (around 45 percent in 1885–1889) were not born in Moscow but were brought from the neighboring provinces where no comparable services existed.[16] The foundling home had its own gynecological and maternity hospital, with separate departments for "legitimate" and "illegitimate" births, and most children born in the second department were given away to the institution.[17]

Infants usually stayed at the foundling home only for several months before being sent to foster peasant families in the countryside. However, during this short stay, almost a half of them died. The home was never meant to handle as many infants as it was receiving in the 1880s, and it suffered from lack of staff, wet nurses, resources, and space. As most of the children were very young (many still had their umbilical cord stump upon admittance), it was extremely difficult to feed them without breastmilk. Furthermore, many children, especially those who came from far away, were given to the home in very poor condition. Such children were usually brought not by their relatives but by professional intermediaries who collected children in the provinces and transported them to Moscow for a fee, often starving them on the way. It is also possible that some negative selection was taking place, and infants who appeared sick or weak at birth were more likely to be abandoned by their families.[18]

Moscow statisticians struggled with how to calculate these infant deaths in the urban demographic statistics. Because so many of the children in the imperial foundling home were born outside of Moscow and brought specifically to be given away, because so many of them died, and because the overall number of births in the city was low, the deaths in the foundling home severely distorted Moscow's infant mortality rate. In addition, considering that abandoned children had only a short stay in the foundling home before being sent to the countryside, some observers asked whether they should be regarded as a part of the urban population at all.[19] As a

A Deadly City for Children

TABLE 9.1. **Infant Mortality in Major European Cities (per 100 Live Births) in 1912**

City	1912	City	1912
Moscow	27.7	Liverpool	12.3
Saint Petersburg	24.8	Dresden	11.9
Warsaw	16.4	Milan	11.9
Breslau	16.3	Marseille	11.4
Cologne	15.2	Birmingham	11.2
Odessa	14.9	Edinburgh	10.5
Vienna	14.9	Paris	10.3
Berlin	14.2	Lyon	9.9
Munich	13.4	Copenhagen	9.8
Leipzig	13.3	London	9.1
Hamburg	13.0	Stockholm	8.6
Manchester	12.5	Amsterdam	6.5

Source: *Otchet of sostoianii narodnogo zdraviia i organizatsii vrachebnoi pomoshchi za 1912 god* (Saint Petersburg, 1914).

result, infant deaths at the foundling home were usually reported separately, and the mortality rates in the city looked quite different depending on whether those deaths were included in the calculation.

In 1891, the reformed regulations of the foundling home restricted admittance to illegitimate or orphaned children and from then on required the documentation about the infants and their mothers. If a mother wanted to keep her name secret, she had to pay a fee of twenty-five rubles—a price too high for most working-class and peasant women. The data on their social background reveals that the largest numbers of abandoned children were coming from women belonging to the peasant estate, most likely migrants from the countryside, with domestic servants being the most common occupation. The foundling home also received the right to trace the mother and to force her to nurse the child held at the institution, unless she had a medical confirmation that she could not breastfeed. Nursing mothers were eligible for a small wage from the foundling home like all other wet nurses. As a result, the annual number of children brought to the institution dropped from more than 16,000 in the late 1880s to below 10,000 in 1894–1895. The survival rate of infants in the foundling home rapidly

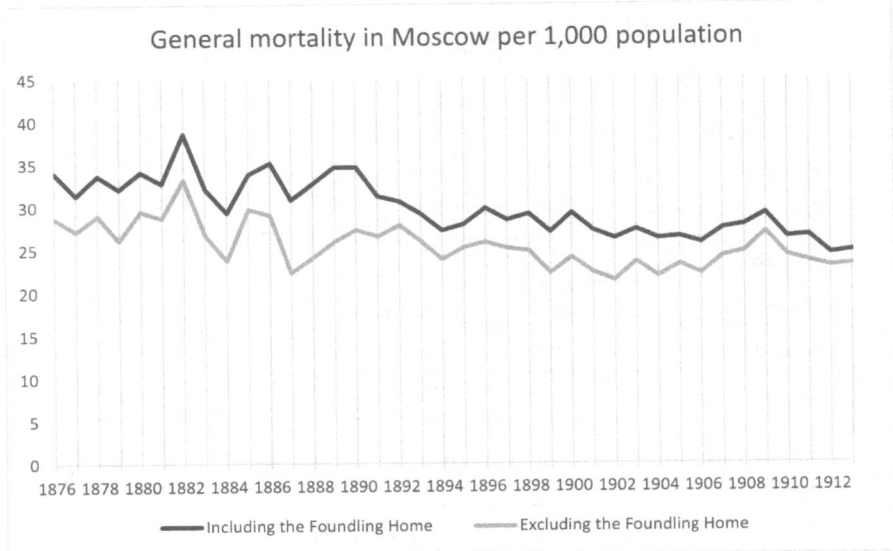

FIGURE 9.1. General mortality in Moscow (per 1,000 population), 1876–1913. Source: *Statisticheskii ezhegodnik g. Moskvy i Moskovskoi gubernii. Vol. 2 Statisticheskie dannye po g. Moskve za 1914–1925 gg.*; P. I. Kurkin, *Estestvennoie dvizheniie naseleniia g. Moskvy i Moskovskoi gubernii: statisticheskii obzor* (Moscow, 1927).

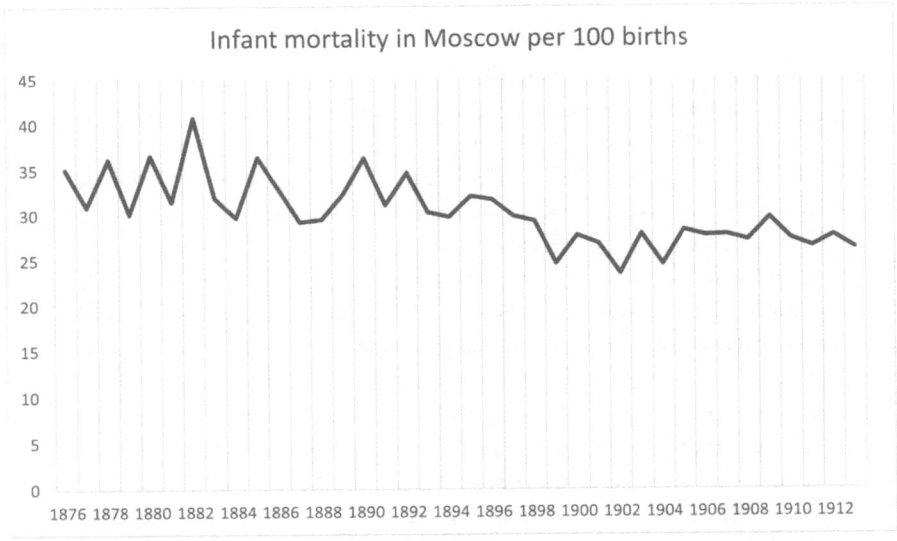

FIGURE 9.2. Infant mortality in Moscow per 100 births 1876–1913 (excluding the children at the foundling home). Source: *Statisticheskii ezhegodnik g. Moskvy i Moskovskoi gubernii. Vol. 2 Statisticheskie dannye po g. Moskve za 1914–1925 gg.*

improved to 70 percent. Simultaneously, Moscow experienced a dramatic decline in overall death rates.[20]

The presence of the Imperial Foundling Home had an important outcome for the perception of child mortality in Moscow. Because in the late 1880s its children accounted for more than one-half of all infant deaths in Moscow, it was possible to see the problem of high infant mortality as primarily the problem of institutional orphanages. The new rules of the foundling home combined with the constant urban growth and increasing birth rate meant that, in the early twentieth century, the institution played a considerably smaller role in Moscow's mortality statistics than in the 1880s. On the eve of the First World War, only one in every five infant deaths happened in the foundling home.[21]

Although the death rate in the foundling home was tremendous, extreme child mortality in Moscow clearly was not and had not been exclusively the problem of orphanages. Even if children in the foundling home were excluded from the calculation, Moscow infant mortality rates remained very high, revealing that there were other factors that made Moscow such a deadly city for children and kept it that way. On the eve of the First World War, Moscow's infant mortality was more than double that of Paris, Hamburg, or Manchester and triple that of London or Stockholm (see table 9.1).

More important, although both children's and overall death rates in Moscow were decreasing at the end of the nineteenth century, that trend was interrupted in the 1900s. In the early twentieth century, exactly when the sanitary reforms in Moscow reached their culmination, when the new sanitary technologies were finally launched, and many new hospitals and outpatient clinics opened, infant mortality stopped declining and so did the overall mortality in the city (see figures 9.1 and 9.2).

The medical causes of infant mortality in Moscow were similar to those in Russian village. The most common cause of death among Moscow children in the first years of life in 1890 (and twenty years later) was gastroenteritis, and such deaths peaked in summer months.[22] Considering that most working-class women in Moscow belonged to the peasant estate and had recently come from the villages of Central Russia, it is likely that their practices of infant care resembled those in the countryside. Like their sisters in the village, they had to return to work soon after birth to provide for themselves and their families. Not having access to breastmilk, at least not when mothers were at work, their children faced similar dangers of malnutrition, water and food contamination, as well as poor hygiene and neglect.

The lack of adequate nutrition and care for poor children in the city was exacerbated by Moscow's notorious overcrowding, which allowed

FIGURE 9.3. Free "corner" apartments provided by the city guardianship of the poor. *Al'bom moskovskikh*. Russian State Library.

historians to call it "the worst-housed city in Europe."[23] In the context of the rapid urban growth and the constant influx of migrants from the countryside, the housing situation in turn-of-the-century Moscow was difficult and barely improving. The growth of residential construction lagged far behind the rise in population, which provoked constant demand for housing and kept rental prices high. Even though the overall city population density per unit of territory was relatively low by contemporary European standards (the same as in London and half that in Paris or Berlin), statistics of the Moscow City Council reveal that the density per housing unit was considerably higher. In 1912, Moscow had 8.7 people per housing unit compared to 4.5 in London, 4.3 in Paris, 4.2 in Vienna, and 3.9 in Berlin. Even Saint Petersburg, which in the Russian view always embodied the evil of the modern city, had 7.4 persons per average housing unit—significantly less than "patriarchal" Moscow.[24]

Although the more well-to-do groups also experienced a shortage of adequate housing in Moscow, the need for housing hit the urban poor the most. In 1908, the average monthly rent for a one-room apartment in Moscow was sixteen rubles—a huge sum for a twenty-ruble worker's budget and completely beyond reach for any single working-class mother. Often unable to afford a flat or even a room, workers opted for so-called cot-and-corner apartments (*koechno-komorochnyie kvartiry*) where renting only a part of the room allowed them to save on housing expenses. The municipal survey in 1899 examined 16,144 of these cot-and-corner apartments, which served as home for almost 175,000 people (which means more than 10 people per apartment), including 40,000 children. The average monthly price of the corner was six rubles, and a cot could be rented for just two.

The housing situation was characterized not only by insufficient space but also by the poor condition of the available premises. The municipal survey reported that 89 percent of the examined apartments were damp, and more than one-third not heated—a remarkable flaw during the frigid Moscow winters.[25] Another survey, carried out by the Russian Technical Society a year earlier, in 1898, found that almost one-half of such "corners" had four or more residents per unit and that kitchens, corridors, and hallways in such apartments were also rented out as sleeping quarters (see figure 9.3). In most of those apartments, residents shared not only rooms but also beds; having one's own cot was a rarity.[26]

In 1899, the researcher Ivan Gornostaev in his report to the Russian Technical Society on the life of working-class children called housing "the most pernicious" environmental condition affecting their life. This report provides many vivid illustrations of children living in such apartments:

"A basement with a Russian oven; the space around the oven is divided into small corners, separated by thin walls that do not reach the low ceiling. One of such corners, 7 arshin [5 meters] in length, 60 vershok [2.7 meters] in width, illuminated with a small window under the ceiling, is shared by two families: the [female] owner with three children and the family F. with three very young children, half-dressed and sharing one cot, resembling a nest with nestlings who open their mouth asking for food."[27]

Although the above quotation clearly reflects how much the housing of the poor offended the sensibilities of their educated observers, it is nevertheless easy to see that such conditions carried real risks to children's health. According to current epidemiological knowledge, overcrowding is associated with an elevated risk of gastroenteritis and diarrhea and respiratory diseases, including tuberculosis, as well as a number of other infectious diseases.[28] This correlates with the main causes of infant death in the late imperial decades. In 1914, almost one-third of all fatalities of Moscow children below five (6,600 deaths) were reported to be caused by gastroenteritis—a slight improvement compared to over 40 percent (6,900 deaths) in 1889. Respiratory diseases (excluding tuberculosis) caused one-quarter of all deaths among children under five in both 1889 and 1914, but the absolute number of their victims actually increased from 4,300 to 5,600. Measles, the fourth major killer in 1914 (after "inborn weakness"), was responsible for 6 percent (about 1,300 child deaths). Other important causes included "diseases of the nervous system," dysentery, and tuberculosis, which each produced between 2 and 3.5 percent of child deaths.[29]

How did Moscow housing arrangements sustain these disease ecologies? Overcrowding, poor ventilation, sharing beds with other household members, and sharing toilets with all other residents of the building facilitated the introduction and spread of infectious diseases, not to mention possible negative effects of the dramatic lack of privacy on the mental health of all residents, both adults and children. Furthermore, the indoor environment and the organization of space in such apartments made it impossible to follow essential rules of personal hygiene and food safety. The residents of cot-and-corner apartments lacked not only indoor plumbing—this was common also in wealthier households. Severely restricted access (or none at all) to kitchens, washrooms, ovens, and stoves and sheer lack of space made it difficult to bathe, wash, do any laundry, cook, boil water or milk—the practices that late imperial physicians and hygienists believed to be so important for preventing infant deaths and that could have interrupted the transmission of gastrointestinal infections. Combined with poverty, inadequate nutrition, or simple hunger as well as the neglect of children whose mothers or both parents had to go to work to provide for their families in the city (and sometimes also in the village), it is perhaps not so surprising that child mortality in Moscow remained so high.

Child Mortality and the Municipal Response

The problem of child mortality reveals an important paradox in the Moscow sanitary project. Throughout the late imperial decades, the deaths of infants and small children were driving up the city's high mortality rates. Both in the 1880s and in the early 1910s, more than half the annual recorded fatalities in the city were the deaths of children under the age of five.[30] Although urban reformers were conscious of these gruesome mortality statistics, it is surprising how little attention was paid to their major cause. The roots of the excessive deaths in early childhood—inadequate nutrition, the lack of care, poor hygiene, overcrowding, child abandonment, and poverty—were largely understood. Possible solutions such as nurseries, hygiene education, improved housing, better medical care, and assistance to poor mothers were identified. The Moscow municipal leaders had no power over employment relations—they could not increase salaries, introduce obligatory maternity leave, or regulate the working conditions of parents—but medical and sanitary provisions, housing regulation, and childcare facilities fell within their responsibility. Yet Moscow had no special program dedicated to the health and care of infants or children similar to the municipality's sanitary, veterinary, food, or school inspections or the inspection of prostitutes.

The fact that this sphere was left out of the municipal public health and sanitary programs is remarkable, considering that both childbirth and schooling underwent strong medicalization. Since the 1880s, the city had invested substantial resources in childbirth care and free birthing institutions. The first four so-called birthing shelters were already established in 1880, and in the following years their number grew to twelve. They were small in size (usually from three to four beds) and run by female midwives and female administrators. In addition, in 1886–1895, the municipality established and financed several birthing wards in the city hospitals. This included a special birthing department for women with syphilis at the Miasnitskaia hospital, which opened in 1888 as part of the municipal campaign against venereal disease and which also resulted in the reform of the sanitary supervision of prostitutes. In the 1890s, municipal childbirth assistance shifted toward birthing houses—bigger hospital-like institutions that were run not by midwives but by obstetricians with a medical degree, usually male (see figures 9.4 and 9.5). In the fifteen years between 1895 and 1910, the number of beds in municipal childbirth clinics increased sixfold. The proportion of childbirths that took place in those institutions grew from 13

FIGURE 9.4. Abrikosova Municipal Birthing House. From *Al'bom zdanii*. National Electronic Library (Russia).

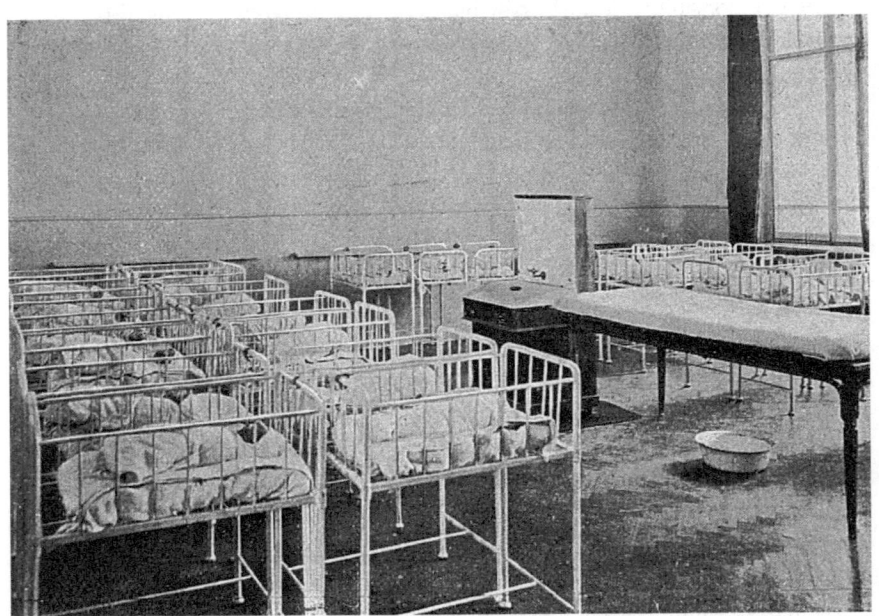

FIGURE 9.5. Abrikosova Municipal Birthing House. Ward for the newborn. From Uspenskii, *Moskva*. National Electronic Library (Russia).

percent in 1890 to 34 percent in 1900 and 60 percent in 1910, showing that medicalized labor was gradually becoming a norm in Moscow.[31]

However, after children left those birthing homes and before they entered school, they fell out of the system of sanitary and public health control unless their parents explicitly sought medical assistance at one of the city hospitals or outpatient clinics. Hospitals, even children's hospitals, did not seem to play any role in addressing the major causes of child deaths in Moscow (diarrheal and respiratory diseases) and, in fact, hardly encountered them at all. Moscow's St. Vladimir Children's Hospital, established in 1876, was praised as a model modern science-based clinic for children and served as a practical school for the professionalizing field of pediatrics. The hospital admitted children under twelve, including infants, and at the end of the nineteenth century served around 2,500 patients a year. Yet hospital records bear little similarity to the mortality statistics in the same age groups. In 1900, in the long list of diagnoses that led to hospitalization (infectious and chronic diseases, tumors, and injuries), only several dozen cases were attributed to diarrheal or respiratory diseases—the conditions that killed thousands of Moscow children that same year.[32] Even though the medical arsenal against those conditions was indeed limited before the advent of antibiotics and oral rehydration therapy, it is still remarkable that the vast majority of such deaths happened outside of hospitals. Moscow's growing and diversifying health-care system could not address the city's exceptional child mortality as most parents did not seek and did not receive medical care. According to some accounts, children below the age of two were even sometimes denied admission to hospitals because of their low survival chances.[33]

The tide began to change only after 1905. By that time, child mortality drew considerable attention from both medical professionals and the larger public. One result of that growing interest was the appearance of voluntary public organizations dedicated to the problem: the All-Russian Union to Combat Child Mortality, with more than five hundred members, including physicians, philanthropists, military officials, and the clergy; the All-Russian Guardianship for the Protection of Motherhood and Infancy; and the Moscow Charitable Society for the Protection of Motherhood. The focus of attention for the last two societies moved from peasant families to poor urban mothers and reflected another important shift in addressing infant care: the treatment of the mother and the child as a unit.[34]

In 1905 the Commission of Municipal Obstetricians and trustees of municipal birthing clinics prepared a report arguing for the extension of medical help around childbirth and additional support to poor mothers in Moscow. The municipal medical council endorsed this report and stated that medical help in the city should encompass also pregnancy and the postpartum period and that it was necessary to establish departments for perinatal diseases and institutions that would specifically support poor women during pregnancy and after birth.[35]

In the following years, municipal leaders established several new modern birthing houses and the in-patient stay of mothers was increased to nine days, giving more opportunities for medical intervention and supervision as well as for the propagation of breastfeeding and hygienic norms of caring for newborns that physicians believed were so important for the prevention of infant mortality. From 1908, Moscow birthing houses started opening so-called consultations—outpatient facilities where parents could get free medical advice on care, health, and nutrition of children under the age of two. In 1910, using the private donation of L. I. Timister, the city opened a perinatal hospital, one of very few institutions in the country designated specifically to care for newborns and their mothers. The Timister hospital had a gynecological and a neonatal outpatient clinic. From 1911, it also had a so-called milk kitchen, specifically aimed at providing adequate infant nutrition to babies that could not be breastfed—usually in the form of sterilized cow's milk diluted with water or an oatmeal decoction in different proportions, depending on the age of the child, which was then put into sterilized bottles and distributed to mothers in need. In 1912, the milk kitchen at the Timister hospital was giving out about six hundred such bottles every day. This was an important but insufficient measure: considering that a baby needed at least six portions per day, it meant that the milk kitchen could serve only about one hundred children simultaneously.[36]

Another angle from which municipal delegates could address the problem of child mortality was the provision of public infant-care and childcare facilities. Such measures were within the responsibility of Russian municipal governments; they were widely recommended by community physicians and had operated in other European metropolises since the middle of the nineteenth century.[37] Moscow did have a system of public nurseries that were popular with poor families. Those nurseries, however, were not considered part of the sanitary project—nor, in fact, of public education—but instead an aspect of charity and were organized by the municipal guardianships of the poor.

These guardianships were a result of the reform of poor relief, which subordinated private initiatives to municipal governments and reorganized the initiatives on the basis of scientific charity. Inspired by the German model, the system of guardianships was a district-based system of poor relief based on voluntary work and funded by donations, members' dues, and an annual municipal subsidy of 40,000 rubles. The initial task of Moscow

A Deadly City for Children

FIGURE 9.6. Children in the municipal nursery of the Presnia district. *Al'bom moskovskikh*. Russian State Library.

guardianships was not to establish institutions but to provide relief in the form of money, materials, or services such as medical care—but from the early 1890s, nurseries became an important institutionalized result of their work.[38]

In 1895, guardianships' childcare facilities housed 340 children. As their task was to care for children while the mothers were at work in factories, shops, or in service, their operation had to adjust to the long working hours of the Moscow labor market. The nurseries were open every day except holidays for about thirteen hours (that is, between 6–7 a.m. and 7–8 p.m.), meaning that children, especially smaller children, spent their entire time awake there. These facilities were usually called *iasli* (literally, "crèche"), which in contemporary Russian means an institution for children under the age of three. In imperial Moscow this word referred to facilities for children under the age of five, at least in an ideal case—as opposed to a kindergarten for children between five and eight. In reality, however, most institutions had no age differentiation, and Moscow nurseries housed children aged between eight months and twelve years (see figure 9.6). Nurseries not only provided food, care, baths, walks, and entertainment for children but sometimes also clothes that the children wore only while they were there. All the services of the nurseries were offered free of charge.[39]

The fact that municipal nurseries were organized by guardianships implied several crucial imprints on their operation. First of all, as guardianships were a form of public charity, the staff at the nurseries were volunteers. They received no salaries and did not have to qualify for their positions. Although upper-class volunteers felt knowledgeable and superior to the working-class mothers who used the nurseries, they had no pedagogical or medical training to support that claim. Second, the guardianships were supposed to be a neighborhood matter, when both members and clients were residents of the same district.[40] This meant there was a huge difference between districts depending on the presence of wealthy donors and active volunteers. In 1903, the prosperous Arbat district had ninety-one free nursery places while the much more populated working-class district of Sushchevo had only forty-five. Third, existing in the borderland between a charity program and a municipal institution, nurseries escaped sanitary regulation and medical supervision. This distinguished them from childcare facilities in other countries, from nurseries in the Russian countryside organized by the zemstvos, and from Moscow primary schools, which since 1889 had been subject to medical control.[41]

Most important, the program of municipal nurseries remained small in scale and could not meet the demand of Muscovites for public childcare, which is evident from the long waiting lists for admission to the existing nurseries. Not only did the municipal program remain much smaller than its involvement in schooling or public health but the expansion of the program slowed down in the early twentieth century, despite rapid urban growth. Increasingly populated city outskirts and suburbs remained particularly disadvantaged. Moscow had 44 municipal nurseries with 1,400 children in 1903 and 50 nurseries with 1,700 children in 1912, while the city population grew by roughly 400,000 over the same period.[42] That year, Moscow had more than 200,000 children of preschool age, and more than 6,000 Moscow infants were abandoned by their parents and given away to the Imperial Foundling Home.[43] Even when combined with the nurseries and shelters organized by factory owners or public organizations, existing facilities could accept only a tiny fraction of little Muscovites in need of public childcare.[44]

With growing urbanization and the employment of women outside of their family and community in the twentieth century, physicians argued for the legislative regulation of mothers' working conditions so that they could safely carry their pregnancy, breastfeed, and adequately care for their children. In 1912, the imperial government passed a law on social insurance that for the first time entitled female employees at large factories to a paid maternity leave in the last two weeks of pregnancy. The same year, the interministerial commission—led by the director of the Medical Council of the Ministry of the Interior, Georgii Rein, a professor of gynecology and obstetrics—developed a plan for the reorganization of medical and sanitary administration in Russia and considered the matter of infant mortality. The project of the national reform developed by the commission, itself already a pale version of the broad social measures advocated by invited experts, proposed the introduction of a six-week paid maternity leave, one hour of paid breaks for breastfeeding employees, and the organization of nurseries by municipalities and zemstvos, covering at least 5 percent of all children under the age of five in the countryside and at least 2 percent in the cities.[45] Although such coverage appears dramatically insufficient to resolve the problem of child mortality facing Russia, it is telling that the available municipal childcare facilities in Moscow could not reach even this moderate goal.

There was also no significant improvement in the housing conditions of poor families with children. Housing in general remained neglected in Moscow's urban policy, although sanitary physicians, guardianships of the poor, and later even specific municipal surveys repeatedly revealed appalling housing conditions of the working population, including families, and drew connections between overcrowding and disease. In 1913, the head of the municipal housing commission, Nikolai Kishkin, openly concluded that "our mortality rates are high because Moscow's attempts to improve the housing of the poor have been very limited and haphazard."[46]

For a long time, however, municipal efforts in the field of housing focused on providing temporary shelter for the numerous Moscow homeless, usually male, who would otherwise sleep on the street or fill the notorious city slums around the Khitrov market, which was the main labor exchange and the first destination of many migrant workers from the village. In 1879, the city opened a "night house," an overnight shelter with 1,300 free beds for night lodging; for more than twenty years it was the only municipal housing project, despite the continuous urban growth and a constant flow of poor migrants from the countryside.

The situation began to change only after the 1905 Revolution. The change was connected with the search for new ways to prevent the unrest and to resolve pressing social problems, but it was definitely helped by the improved financial situation of the municipality thanks to the profits from municipal enterprises and generous private donations as well as increased subsidies from the central government. Furthermore, a combination of outbreaks of several infectious diseases (cholera among them) that hit the city in 1908–1910 again highlighted the vulnerability of the urban poor to epidemics and contributed to the urgency of the housing problem.

In 1906–1909, six new night shelters opened in Moscow with 4,350 beds altogether. This was an impressive addition, but it was still insufficient; the homeless population of Moscow was estimated at 15,000.[47] In addition to temporary lodging for the homeless, in the late 1900s the Moscow municipality started providing cheap municipal apartments to the poor. In 1909, the city established two houses (Solodovnikov houses) with apartments for two thousand people. These houses offered small furnished apartments for about five rubles a month with shared kitchens, bathrooms, and (importantly) public nurseries for children living in those buildings. There were also houses with free apartments for specific categories of the urban population: Bakhrushin House for two thousand widows and single women coming to Moscow to study, Boev House for families with orphans, Tretyakov House for widows and orphans of Russian artists. All these housing initiatives, however, could help only a small group of those living in overcrowded apartments.[48]

One aspect of this dramatic and barely improving housing situation in Moscow was financial. According to the municipal survey of cot-and-corner apartments, their existing price was the maximum their residents could pay. This meant that the owners of such apartments had little motivation to invest in their property, because in most cases they could not raise the rent, especially in less desirable locations such as basements or in buildings without access to the sewerage system. The problem needed municipal involvement, but housing projects for the poor were costly and could not easily serve as symbols of modernity, progress, or service to the entire urban community, which helped recruit support for other expensive municipal projects such as the abattoir and the sewerage system. As a result, for a long time the funding for municipal housing initiatives came from private donations (the pattern commemorated in the names of the housing projects that usually bore the name of the donor). Those donations were sometimes very considerable, such as the six-million-ruble donation by a merchant called Gavrila Solodovnikov, which financed the construction of the Solodovnikov houses with cheap apartments. All the affordable housing projects listed above were built with private donations, as were at least three of six municipal night shelters. However, when using private donations, the municipality was bound by the specific conditions of the donor and had to find an agreement with the executors of the will, which sometimes meant that the money could not be used efficiently.

Another obstacle was simply the lack of authority to regulate private housing. Contemporaries echoed each other in their complaints about the limits of municipal housing control in Russia.[49] In the 1900s, the Moscow City Council twice sent a petition to the Ministry of the Interior asking for the right to close flophouses around the Khitrov market in view of their unsanitary condition, but the eventual response was that the widening of the rights of the city government could take place only after the entire municipal statute had been changed. In the 1910s, two more attempts to improve the situation at the Khitrov market were halted or delayed by the imperial administration and could not be implemented before the outbreak of the war.[50]

Even though the closing of the flophouses failed, municipal leaders were well aware that restrictions alone would hardly improve the housing situation of the poor. They reasonably feared that if the flophouses and cot-and-corner apartments were closed, their residents would become homeless unless alternative forms of affordable housing were provided.[51] Such alternatives did not appear until the eve of the First World War. In 1911, a municipal deputy and engineer called Emmanuil Albrecht developed a plan to build affordable housing financed through municipal loans—the strategy that had been used for building sanitary technologies since the times of Alekseev. Albrecht believed that the rents in such houses would eventually cover the cost of their construction and exploitation. He proposed building sixty five-story houses with about 9,500 cheap apartments, which, according to his estimation, would cost between eight and nine million rubles. This project would not only provide housing to those who needed it (or rather, a wealthier part of that group), but it would give the municipality a chance to push for stricter control over the cot-and-corner apartments without the risk of making their residents homeless. Although the plan received the support of the city housing commission, in it got stuck in municipal discussions and was not implemented in the remaining years of the imperial period.[52]

This was the most thorough housing project of the Moscow municipality, and even though unrealized, it reveals how municipal leaders envisioned the proper life of poor families. On one hand, the houses would have had modern sanitary amenities such as running water, sewerage, laundries, and disinfection chambers as well as social infrastructure in the form of daycare facilities for children. On the other hand, Albrecht calculated that the project would house 39,000 people—that is, on average, four people per apartment. Each apartment would consist of only one room measuring between ten and twenty square meters, where the entire family would live without any separation of genders or generations. The kitchens would be shared too, with one kitchen for fifteen apartments or sixty people. Even if the project was sanitary and modern, it is telling how little space and privacy this most daring housing reform in imperial Moscow allowed for.

The Moscow sanitary project failed to address the most important health problem of urban childhood—that is, the extreme death rates among infants and small children, and this explains why the overall mortality in Moscow remained so high and why it was declining so slowly. In the early twentieth century, overall fatalities from infectious diseases were high in Moscow, but they were comparable to those reported in other European cities, especially in Southern and Central Europe.[53] It was the dramatic infant mortality that consistently set Moscow apart and that was responsible for its status as the "deadliest metropolis" in Europe.

It is often more difficult to see the reason why political actors did not address a specific problem than the reason they did. For the implemented sanitary reforms there is abundant documentation left that allows us to reconstruct the motivations behind them, the explanations provided, the goals set, even if not achieved, and the paths not taken. For the reforms that did not take place and were not even planned, no such material exists, leaving space for more or less informed guesses.

One possible reason the government of late imperial Moscow had no program to tackle child mortality was that it was not seen as a sphere demanding targeted municipal intervention during the Alekseevan period, a crucial time for the Moscow sanitary project when its key directions and formats were determined. At the turn of the 1890s, the number of children born in Moscow was relatively small, and the large proportion of infant deaths in the city occurred in the Imperial Foundling Home, which was not controlled by the municipality. Furthermore, the home was itself being reformed. In that situation, it was possible to hope that infant mortality would diminish because of the improved arrangements at the foundling home and the broader public health and measures aimed at the entire urban community—such as the sanitary inspection, the sewage system and the water pipe, better access to medical services, and the poor relief that were all being established or expanded under Alekseev. There were also some more specific measures that could be framed as promoting the health of Moscow's newborns, such as maternity shelters and the program to control venereal disease, which was viewed by many as an important cause of congenital defects and infant deaths. In fact, it is likely that some of these interventions were indeed having an effect as infant mortality did gradually decline until the early 1900s, and this positive trend made the overall problem appear less urgent.

However, these measures turned out to be insufficient in the long run, especially as the city continued to grow demographically and to expand spatially. In the early twentieth century, rates of infant mortality stopped declining, and the absolute number of children dying and their proportion in the total fatalities in Moscow actually increased. This, combined with the heightened public attention to child mortality in Russia in the early twentieth century, the professionalization of pediatrics and obstetrics, made the problem more acute and visible. The appearance in 1902 of the Municipal Committee of Obstetricians, which included all obstetricians and female directors of maternity shelters, facilitated the lobbying of specific measures on infant care within municipal structures. The resulting policy changes—such as the expansion of maternal, neonatal, and pediatric care and the opening of milk kitchens and "consultations" for mothers—appeared only in the last decade of the empire. Although these new institutions did impress foreign observers such as Charles-Edward Amory Winslow, they remained too small in scale and too unevenly distributed across the city space for their results to be felt in the remaining years before the Bolshevik revolution.

In the 1920s, the new Soviet government would make a concerted effort to address the problem of both housing and child mortality in their new capital. The Soviet policy on infant welfare in Moscow absorbed many formats that had been developed and tested before the revolution. Outpatient clinics, maternity homes, "consultations," milk kitchens providing free food for infants, and public nurseries would become typical features of Soviet Moscow. The scale and the impact, however, were different. The introduction of the eight-hour working day, maternity leave before and after childbirth, financial aid to new parents, regulated breaks for breastfeeding mothers, the wide organization of nurseries, an aggressive propaganda campaign on hygiene and infant care, and home visits by physicians and nurses, as well as legalized abortions, all led to a rapid decrease in child mortality in Moscow. By the mid-1920s, the number of infant deaths had dropped to under 19 per 100 births, pulling down also the general mortality rates in the city. These falling numbers would be used as powerful justification for the success of the Soviet system.[54]

The housing question proved more difficult to tackle because the war, the revolution, and the relocation of the central government from Saint Petersburg to Moscow turned the city housing problem into a veritable crisis. Unable to provide more and better apartments, Soviet authorities focused on the expropriation and redistribution of the existing housing stock, resettling the poor from barracks and flophouses to the homes of the wealthy. This measure, although obviously highly unpopular among the former owners, indeed helped to ease the housing need among the working-class population, even if only temporarily. With an average rate of two or three people per room, the arrangements in communal apartments provided more space and privacy not only compared to the flophouses but also to the most progressive working-class housing projects of the imperial period.[55] As the city continued to grow and housing provisions continued to fall behind, the communal apartment would remain the typical residential arrangement in Moscow for decades to come, until the mass housing construction in the 1960s allowed the majority of Muscovites to get a private home.

CHAPTER 10

Healthy Schools in a Deadly City

The late imperial period saw considerable changes in childhood policy and experience in Russia, and perhaps nowhere were these changes more apparent than in schooling. The old norms of adults' unquestionable authority over children, their oppressive treatment, and mechanistic teaching were challenged by changing concepts of childhood and new ideas of upbringing with humanistic, child-centered, and communicative approaches. The post-reform decades recorded an unprecedented expansion of primary schools. According to the statistics of the Ministry of Education and the Holy Synod, the number of schools (including municipal, zemstvo, and church parish schools) grew from about 8,000 in 1856 to over 100,000 in 1911; the number of pupils increased from 450,000 to 6.6 million over the same period. Of course there were substantial regional variations in school availability, and the expansion of primary education had to catch up with the population growth of the early twentieth century, but it is nevertheless clear that Russia was gradually moving toward a schooled society.[1]

The focus of existing historical research on Russian schools has been on the content and form of the teaching, the political drivers and intellectual foundations of the educational reform, and the impact it had.[2] However, schools were also a physical and social environment in which children spent a large part of their day and which affected their academic performance, their bodies, their health, and their lived experience of schooling. The environment of Russian schools was transforming, and this transformation was shaped and motivated by new medical knowledge and the rise of public health and sanitary reforms.

In the 1870s, physicians and hygienists joined pedagogues and educators in the debates about schooling and its needs. Educational reformers recognized that school could shape not only the minds, knowledge, and morality of the pupils but also their bodies and physical development, and that the two spheres were in fact tightly interconnected. This meant that, although the content and style of teaching remained the primary concern, there were now new variables that could determine the results of schooling: the material environment of schools, the temporal and spatial organization of the educational process and its ability to accommodate, adjust to, harm, or change the pupils' bodies.

For public hygienists and community physicians, schools presented an excellent source of information and an object of medical statistical research. Few other institutions offered such a possibility to observe and study patterns of health and disease. The fact that in post-reform Russia the development of community medicine and, to a substantial degree, the expansion of schooling were managed by the same local self-government bodies—zemstvos and municipalities—helped the intellectual exchange between the two spheres and opened the way to some synergy of practical efforts.

In this chapter I will examine Moscow municipal primary schools from the angle of school hygiene and urban public health to see how sanitary reforms affected the school environment and the experience of children. There are several aspects that make the study of Moscow schools particularly interesting. First, among Moscow's child population, it was the pupils of municipal schools and their health that received the most attention, and there was a visible effort to offer them better medical care and a healthier environment. This attention appears in sharp contrast to the laissez-faire policy toward the health of infants and small children, the groups with much higher mortality rates. Second, Moscow was very early, both in Russia and internationally, to institutionalize medical control in municipal schools through the introduction of school sanitary inspection in 1889 and the establishment of school outpatient clinics. Furthermore, both municipal leaders and municipal school physicians showed a remarkable commitment to a more inclusive gender policy, promoting school education for girls and the right to medical practice for women. Finally, Moscow, unlike many smaller towns or zemstvos, had enough financial, social, and infrastructural resources to actually implement at least some of the expert recommendations and to translate scientific ideas into practice.

School Environment, Medicine, and Children's Well-Being

"Nowadays there is a widespread opinion that the present organization of schools harms the health of children," wrote Friedrich Erismann in 1870 in his book about the influence of schools on the development of myopia in children.[3] This was the "first" book in several respects—the first book that Erismann, born and trained in Switzerland, published in Russia, the first book in which he moved beyond his initial specialization (ophthalmology) into the domain of public hygiene, and the first book that applied the ideas of Western European hygienists to the study of Russian schools.

Erismann's interest in schools and their influence on eyesight was not particularly innovative in itself. In the 1860s, several European physicians studied the adverse impact that schools had on pupils' health.[4] In 1869, on the request of the Prussian minister of education, Rudolf Virchow brought these accounts together in his report *Ueber gewisse die Gesundheit benachteiligende Einflüsse der Schulen* (Some adverse effects of the schools on health).[5] The report inspired a significant resonance in Russia as the deputy minister of public instruction ordered it to be translated and published in the ministry's official journal. The journal editors noted that Virchow's valuable observations could be of limited practical interest in Russia where the primary concern was the lack of schools rather than their negative impact, but by 1870, Virchow's report had already appeared in Russian in two different translations.[6] These publications, together with Erismann's book, signaled the beginning of school hygiene in Russia, which would then develop and grow more institutionalized in the following two decades.

So how exactly did nineteenth-century schools in the view of medical professionals, harm the health of their pupils? The answers Russian hygienists gave to this question were largely similar to those of their colleagues in Western Europe and America. These answers were rooted in the belief that social and environmental factors had a profound impact on human health—the belief that dominated European medicine in the mid-nineteenth century and persisted in Russia also after the acceptance of the germ theory of disease. On the other hand, these answers embraced the developmental physiology of childhood and were based on the idea that children's needs, abilities, and disease profile differed from those of adults and changed with age.[7]

These two sets of assumptions resulted in attempts to identify specific childhood- and school-related diseases and methods for their prevention through the reorganization of the school environment. The problem of school-related diseases was complex and multilayered. First, there were diseases and physiological, ophthalmological, or neurological conditions that developed directly under the influence of school and schooling. In the view of Russian hygienists, these were headaches, anemia, myopia, strabismus, hysteria, or neurasthenia as well as injuries and wounds that occurred because of inadequate organization and care in schools. There were diseases that were not directly caused by schooling but for which going to school was believed to be a predisposition: diseases of eyes and eyelids, digestive disorders, and tuberculosis. Finally, there were acute infectious diseases that found fertile ground in schools.[8]

One specific concern in Russia was the weakness and underdevelopment of pupils' chests and ribcages, at the time considered a predisposition to consumption and other diseases of the lungs. In 1881, a zemstvo sanitary physician and veterinarian (later one of the founding fathers of the Moscow abattoir), Valentin Nagorskii, examined pupils of the Saint Petersburg zemstvo district and found out that in their physical development, including height, weight, and especially chest girth, they yielded not only to pupils from Western European countries but also to their coevals employed at Russian factories. Was a school, Nagorskii wondered, more dangerous for children's health than a factory? Given the existing hygienic state of schools, he suggested, it was perhaps a blessing that only a minority of children were attending educational institutions, because the benefit for their intellectual development could hardly make up for the damage done to their health.[9]

Medical experts indicated many features of Russian schools that had a negative impact on children's health. School furniture was one. Erismann noted that in Saint Petersburg, where he conducted his first survey on school hygiene, "very little attention was paid to the height of pupils, therefore 10-year old boys often work at the same desks as 20-year old men, so they cannot reach the floor with their feet and, because of the extremely high position of desks, are forced to lift their shoulders so much that their necks become completely invisible."[10] In addition, the organization of space and furniture at school caused constant inconvenience and discomfort to pupils, forcing them to move, turn, and fidget, which undermined their concentration. Although teachers attributed children's lack of attention and inability to sit straight to their negligence, inadvertence, and bad manners, hygienists argued that those problems were a result of the inadequate school environment. Desks and benches that were either too big or too small, the pupils' inability to adjust their position, the lack of backrests and footboards were easily identifiable problems and also relatively simple to amend.

Other widely acknowledged problems were more difficult to tackle because their resolution required a complete reconstruction and relocation of existing schools. These included insufficient lighting in classrooms, poor ventilation, too much dust and too little oxygen in the air, dampness, inadequate heating, badly organized washrooms, toilets, or privies. The standard for school buildings that hygienists were arguing for was not easy to meet. Schools had to be spacious, dry, well lit, and well heated, with several rooms, a teacher's apartment, and a yard. Erismann's ideal classroom was a seventy-square-meter room with at least four-meter-high ceilings, a window on the left side, oak parquetry, diffused lighting, and independent systems of heating and ventilation. It was meant for a class of thirty-six, or eighteen double desks arranged in three rows. The size of the room was supposed to allow all pupils to see the blackboard and to hear the teacher's voice without it being confused by any echo. Instead of standard flat school desks, Erismann proposed using slanted desks with an incline of between twelve and fourteen degrees (the design later known as Erismann's desk), which he believed to be the most ergonomic and beneficial for pupils' posture and sight.[11]

The reality, of course, fell behind those hygienic norms. A sanitary engineer, Illarion Pavlov, observed in 1886 that "although school hygiene is sufficiently developed, although it provides general rational rules of classroom size, lighting, heating, ventilation, etc., until now hygiene existed on its own and reality on its own." In his view, an important reason was the failure of engineers and technicians to provide an essential link between the two spheres and to produce projects that considered both the norms of hygiene and the resources of community schools.[12]

The material organization of schools was, however, not the only concern of hygienists. The aspects of school life that hygienists saw as being in the domain of their influence, responsibility, and intervention were surprisingly numerous. If the focus on school furniture, organization of classrooms, ventilation, and sufficient lighting reflected the preoccupation with schools as a place, another set of issues dealt with schooling as a process. This concern with the process of schooling—and, especially, the exhaustion of pupils—was again something that Russian hygienists shared with their colleagues in Western Europe or America, but some aspects of schooling seemed to generate particular attention in Russia.[13]

One such issue was school discipline and punishment. In this question, the positions of Russian hygienists and reform-minded pedagogues were unanimous. Nikolai Korf, in his famous and influential handbook for teachers, *Russkaia nachal'naia shkola* (*The Russian Primary School*, 1870), which by the turn of the twentieth century had gone through two dozen editions, called for abandoning the "old" military-style school discipline based on rods, fear, oppression, and boredom and argued that only a warm and loving attitude toward children could lead to successful learning.[14]

This view soon became a widespread teaching philosophy. If for progressive educators the rods of the "old" school were pedagogically ineffective, for hygienists they were unhealthy. The hygienists opposed not only all obvious forms of corporal punishment such as flogging but any disciplinary measures that involved the body—flicks and slaps, hitting pupils with a ruler, making them kneel or stand, leaving them without a meal, and so on. The only acceptable form of punishment was to deprive a pupil of some pleasure—for example, a game—but, as one physician admitted, "there are very few pleasures in school life."[15]

Curriculum was another sphere of the schooling process in which hygienists wanted to interfere. They insisted on adjusting schooling to the psychological development of children and easing the strain it caused on their mental and physical health.[16] Whereas pedagogues and educators argued for the expansion of schooling, for the possibility to teach more subjects and more classes to more people, especially in primary schools, physicians proposed limiting it. Hygiene, Erismann wrote, "should require the simplification and reduction of school curricula, that is the decrease in the number of subjects, in the number of lessons, especially among younger pupils, the decrease in the quantity of homework and preparation. It is unacceptable that a fourteen-year old child spends all day with books, at school or at home, and that he does not have time for outdoor movements, for games or any other physical activity."[17]

This emphasis on game and the health value of playing was another distinctive feature of Russian medical discourse on the process of schooling. To minimize the negative effect of schooling and to keep the balance between the development of mind and body, hygienists prescribed sufficient sleep and food, long walks, and physical exercise. Among possible types of physical activity, it was not structured and disciplined training or gymnastics but playing outdoors that was seen as the healthiest and the most suitable option for schoolchildren. In Erismann's words, "our children play very little, and our urban children do not even know how to play. This phenomenon at first seems very strange and its roots are hidden in many natural and practical [estestvennykh i bytovykh] circumstances of our home country.... Children need to play; for any child a game is a necessary condition of life and normal development. If our society gets used to the idea that active games should not take place in closed premises but, if possible, outdoors, this would create a base for the proper physical development of our younger generations."[18]

Although the focus of such recommendations was not directly on school as a place, they did have an important spatial dimension. Physical education, games, and outdoor activities required sports rooms, playgrounds, and gardens; meals required facilities where they could be cooked and consumed. Such recommendations, therefore, had direct implications for how a healthy school should be built and where it should be located.

There were two important consequences of framing the field of school hygiene so broadly. First, school hygiene provided a language and tools to criticize schools, even the most "progressive" municipal and zemstvo schools, from the position of a child's experience (however misinterpreted by hygienists), rather than academic achievement. This perspective offered an alternative to the excitement about the rapid spread of schooling in the post-reform Russian society. Hygienists were far from denying the need and the value of mass education, but they warned that it had its cost. Schooling—even if it promised personal development, social mobility, and liberation in the future—still required restraining the body and the freedom of a pupil, condemning him or her to monotonous days in an uncomfortable and unnatural position, often hungry and cold, and at risk of getting a chronic or contagious disease. Therefore, physicians argued, the classroom experience should be minimized, diversified, and compensated for with sufficient time outside of school and away from the educational process.

In their own narrative, theoreticians of school hygiene saw themselves as protectors of pupils and their bodies against the coercion of the educational system. The question remains, however, whether the lived experience of children outside of schools was any better or freer than at school. Ben Eklof's research on Russian rural schools reveals the enthusiasm with which children went to school and the affection they retained for schooling. Eklof also shows that the new child-centered and humanistic pedagogy encountered resistance within the families—parents thought that children were treated too leniently at school, that school was spoiling them, and they encouraged teachers not to spare the rod.[19] Given the harsh family mores among Russia's laboring population, common domestic violence, authoritarian parental power, and cruel child-rearing practices in villages and cities alike, the role of schools in children's physical and psychological health was both restraining and liberating. School could be not only the source of disciplining, physical and mental exhaustion, chronic and contagious disease but also an escape from widespread violence and oppression and an alternative to hard work at a factory, in a workshop, or in the household—that is, a healthier and safer space for a child's body.

On the other hand, the widest possible delineation of the domain of school hygiene also served as a powerful and, in the Russian case, sometimes effective justification for the physicians' claim to greater authority in matters of education and more control over the operation of schools.

Already in 1877, Erismann welcomed such a medicalization of schooling:

> The beneficial and desirable development of school affairs in the interests of students will only be possible if teachers and directors of educational institutions take the question of school sanitary conditions seriously and if physicians with special education in hygiene receive a significant influence over the organization of school curricula and over the lessons themselves. In other words, the physical and mental well-being of the youth urgently requires the organization of sanitary control over the state and private educational institutions and the active involvement of hygienists in the decision-making of school councils.[20]

A decade later, another hygienist, Aleksandr Virenius, argued that a school doctor should work closely with teachers and that it is only through their collaborative efforts, through the integration of pedagogy and school hygiene, that true progress in education could be achieved. Unlike Erismann, Virenius believed that hygienists not only should have control over the school buildings and the structure, organization, and length of lessons but should directly interfere in the pedagogical questions and the form and content of teaching:

> In resolving almost all questions concerning the pupil, the matter cannot avoid an opinion, advice, or a recommendation of a physician. The evaluation of abilities and a character of a child, finding the measures to correction or improvement, the distribution of lessons according to the mental abilities of children, the adequate development of the intellect and morality, etc., all of this should be guided by the laws of physiology and psychology—that is by the recommendations of one of the representatives of these fields of knowledge, the physician. In general, the participation of the physician in purely pedagogical questions is necessary and it is now gradually being required in civilized countries.[21]

Similar claims for authority and greater control over schooling were shared by many physicians across the world. In Moscow, physicians saw some of their ambitions come true with the organization of the municipal medical inspection of schools.

Moscow Primary Schools in Transformation and the Emergence of Medical Control

A researcher of school hygiene and sanitary control in Moscow would be surprised by two aspects. The first is the early institutionalization of medical inspection at schools. Although Russian physicians and Moscow municipal leaders often portrayed Moscow as backward and in desperate need of catching up with more "advanced" societies in the West, school hygiene was one sphere where Moscow did not fall behind. Physicians in Western Europe and North America were actively exposing the health problems of the school environment and the schooling process but until the 1890s and 1900s they had little control in educational institutions. When physicians were finally allowed into schools, it was not to transform the school environment but to trace cases of infectious disease or to provide a medical explanation to differences in academic achievement.[22] In Moscow, medical supervision of schools was institutionalized as early as in 1889, and it was precisely the broad concerns about the relationship between the school environment and children's health that defined the agenda of school physicians in the city.

The second aspect is the apparent lack of resistance and conflict over the institutionalization of school medical control. In other countries, school officials and teachers resisted the intervention of physicians, seeing it as a threat to their own authority.[23] In Moscow (at least according to the preserved archival sources), this did not seem to be the case. The initiative to establish medical supervision in fact came from the city schools themselves and then quickly found generous support in the municipality.

Primary education was one of the early municipal projects in Moscow. In the late 1860s, Moscow, then a city with a population of about 400,000, had only 13 public elementary schools for boys. Overseen by the Ministry of Public Instruction, they were subsidized by the Moscow City Council. In 1867, to balance this gender disproportion, the city government opened five girls' schools, and those became Moscow's first municipal schools.[24] The 1871 report by the inspector of popular schools from the Ministry of Public Instruction gives a picture of how the first municipal primary schools were organized and operated—five for girls and one for boys, opened in 1870. The boys' school had 126 pupils and employed six teachers and two priests. The girls' schools were somewhat smaller in size: each of them had about 100 pupils, one priest, and three or four teachers, usually female. In addition, each school also had a (female) trustee (*popechitel'nitsa*), responsible for the supervision and administration of the school. Pupils were divided into three grades according to their abilities and studied reading and writing, grammar, basic Russian history and geography (*mirovedenie*), arithmetic, religious instruction (*Zakon Bozhii*), as well as singing and mechanical drawing (*cherchenie*).[25]

Municipal primary education was not free, but the tuition fee was set at only three rubles per year—compared, for example, to more than two hundred rubles per year at a private elementary school in Moscow. Even that sum was apparently too high for many families, however, and as the

TABLE 10.1. **Expansion of Municipal Schools in Moscow**

	1869– 1870	1879– 1880	1889– 1890	1899– 1900	1909– 1910
Number of schools	5	40	81	150	288
Number of classes	12	119	267	501	1,170
Number of pupils	331	4,138	1,0461	19,853	43,532

Source: Verner, *Sovremennoe khoziaistvo*.

report reveals, a large proportion of pupils (sometimes more than one-half) studied free of charge.[26] The tuition fees were not meant to pay for school expenses, which were covered by generous municipal funding but, rather, allowed schools to accumulate some additional funds; perhaps, this could explain the lenience in collecting the fees.

In 1882, Moscow had fifty-five municipal elementary schools, including 26 schools for girls, 25 for boys, and four for both sexes together. That year the Moscow city councilors declared the systematic expansion of primary education to be a priority and set about to establish ten new primary schools a year. After this, the increase in the number of schools continued rapidly; by 1910, Moscow had 288 schools (see table 10.1). Moscow also took steps to develop secondary education: in 1885, the first two municipal secondary schools for girls were opened, to be joined by a secondary school for boys several years later—notably, girls' schooling was again taking priority. However, the number of municipal secondary schools remained very small (seven for boys, eight for girls in 1911–1912), and municipal efforts were concentrated in primary education.

In 1909, the Moscow City Council members adopted a course toward universal primary schooling. At the same time, the three-ruble tuition fee was abolished and the length of study at Moscow municipal schools was increased from three to four years. By 1911–1912, Moscow already had 312 primary schools and all of them had successfully switched to a four-year course. The financial side of this project was helped by a governmental subsidy, resulting from the State Duma's decree on sponsoring public education. The Moscow City Council petitioned the Ministry of Public Instruction to make primary education in Moscow obligatory; but the ministry's position was that the introduction of obligatory primary education could only follow revision of the general law.[27]

Primary education in Moscow was separate for boys and girls, although a small number of mixed schools existed between 1879 and 1893. The goal of keeping the gender balance, which had been behind the municipal intervention in public schooling in the 1860s, never disappeared, as municipal leaders remained committed to promoting both boys' and girls' education. Despite the general bias against girls' education in Russian society and the stronger motivation for boys to finish elementary school (it allowed them to reduce the term of mandatory military service in the future), proportions of male and female students remained, respectively, at about 52 percent and 48 percent, but there were more schools for girls than for boys because, as the former tended to be somewhat smaller in size. Furthermore, girls' schools had predominantly female teachers and exclusively female trustees. The existence of trustees—responsible for administration, maintenance, teaching arrangements, and personnel decisions at their respective schools—was a peculiar policy of Moscow; it was different, for example, from that in Saint Petersburg, where several schools were managed by one district trustee, usually male. The practice of having only female trustees for girls' schools meant that more than half of Moscow schools were managed by women. In addition, school trustees were often consulted and invited to attend the meetings of the School Committee of the Moscow City Council, allowing women to take an active role in shaping public education in Moscow.[28]

Who attended those municipal schools and how? The 1901–1902 report of Moscow primary schools gives some idea of student profile and attendance. That year, the city had 176 primary schools, with 11,824 male and 10,999 female students. Those pupils were rather unevenly distributed across the school grades. The most common size of a first grade was between 45 and 55 pupils—compared to between 35 and 55 in second grade and between 15 and 35 in third grade. This suggests that many pupils withdrew without finishing a course (this trend was particularly noticeable in girls' schools). The absolute majority of pupils at municipal primary schools were the children of the urban poor. About 55 percent of all pupils belonged to the peasant estate (this group, no doubt, counting many migrant workers at factories and workshops); one-third were from the lower urban groups and craftsmen (*meshchane i tsekhovye*), 5 percent were "soldiers' children," and only 6.5 percent came from families of merchants, priests, honorable citizens, and other privileged groups.[29]

Although in the late 1880s, school hygiene was already an established field in Russia and although urban public health was at the top of the municipal agenda at the time, the initiative to subject expanding municipal schools to medical control came not from physicians but from the school administrators. In October 1887, Nikolai Richter, the trustee of the boys' elementary school in the Prechistinskaia district, suggested that the Moscow municipal board appoint a sanitary physician to his institution. "Concerned with the sanitary state of the school and pupils," Richter consulted his acquaintance Nikolai Mikhailov, a former zemstvo sanitary physician, who agreed to perform medical and sanitary control at his school—and, remarkably, without any compensation for his work.[30]

Mikhailov was, in fact, an experienced sanitary physician with a name of some renown in school hygiene. As a sanitary physician of the Moscow zemstvo, he conducted research and published on the physical development and the morbidity of pupils at rural schools as well as on the sanitary conditions of educational institutions.[31] Using his experience of inspecting rural schools, Mikhailov prepared a draft program of responsibilities of school sanitary physicians, which Richter attached to his letter—another telling example of the connections between the Moscow zemstvo and the Moscow municipality in the field of sanitary reforms. The program included medical examination of all children entering schools, smallpox vaccination, biannual measurement of children's growth, control of their health, quarantining and providing basic medical care, issuing certificates of recovery, as well as inspection of sanitary conditions at schools and disinfection.[32] Although Richter's stated goal was to get the board members' approval for his innovative practice, it is plausible that the actual purpose of the letter—and definitely its eventual result—was to attract attention to the matters of health and hygiene at schools.

Richter's letter was received well by the board and raised the question of organizing systematic medical inspection of the city schools. To discuss the matter, the teaching commission of the municipal board convened a meeting of school trustees (both male and female), municipal representatives, and physicians with experience in inspecting children's health. The participants agreed that the establishment of medical supervision at schools would be a good way to prevent the spread of contagious diseases and the development of chronic diseases and that it would be easier, cheaper, and more convenient to organize such control in a centralized manner. It was proposed to hire six physicians, whose work would be compensated by the municipality from existing school tuition fees. Views differed, however, as to the exact remuneration of physicians—some suggested their salary should be 780 rubles per year, like that of physicians at municipal hospitals; others thought it should be 1,080 rubles, like that of sanitary physicians at the night shelters.[33]

The purpose of the school medical inspection was not seen to be cure or therapy but, rather, monitoring and prevention. The pupils found sick would be referred to the city hospitals for treatment. The participants of the meeting generally supported the program proposed by Mikhailov but added that "because of the novelty of this activity for Moscow, the detailed regulation of the tasks of a [school] physician is impossible: it should be left to experience."[34]

The Moscow City Council members approved the plan and, in fact, agreed to allocate more funding to it than had been initially requested. The salary levels were set at 1,080 rubles per year for five regular physicians and 1,500 rubles for the chief physician. The shape of the sanitary inspection at schools was decided not by state bureaucrats or medical scientists but by local teachers, school administrators, and public health practitioners, who, although perhaps lacking competence in the scholarly debates on child physiology and psychology, had a good understanding of the actual practice of schooling in Moscow and of children's experience in schools.

The success of this project was largely contingent. It was the result of an initiative of a concerned school trustee and his personal connection with a motivated physician who was apparently ready to volunteer for the cause. On the other hand, the initiative clearly came at the right moment. This was exactly the time when the municipality under Alekseev was actively expanding its authority in public health and was specifically developing the format of sanitary inspection. It was also the period when decision-making routes and times were short, when municipal leaders were often in direct conversation with experts, and when initiatives that fitted into Alekseev's vision of municipal development were likely to receive quick approval and financial support. Finally, even though it seems Erismann was not directly involved in this proposal, it is clear that the initiative resonated with the work of Moscow's most influential hygienist, who regularly acted as an advisor to the Moscow government and whose research had created a fertile ground for attempts to subject schools to medical control.

Moscow School Physicians, Their Work, and a Healthy Environment for Schoolchildren

The practical activity of school physicians in Russia has largely been overlooked by historians or dismissed as a failure. Andy Byford writes that "the hygienists' conceptualization of the school doctor remained only an unrealized ideal. In practice, Russian school doctors were ordinary general practitioners with only a formal link to a few schools in their local area. . . . Only very occasionally and entirely as a matter of the individual doctor's personal initiative would systematic studies of, say, the student's eyesight, the quality of air in classrooms, or the adequacy of lighting in a school, be carried out. In other words, issues of 'school hygiene' were not at all a regular part of doctors' job description."[35] For Byford, it was psychology and psychiatry that served as a link between medicine and education and promised to empower school physicians, particularly when dealing with "unteachable" or "abnormal" children.[36] An analysis of medical control at Moscow municipal elementary schools with its institutionalized and systematic inspection already in the last decade of the nineteenth century—that is, before the rise of child psychopathology—offers an important correction of this view and a different interpretation of the role that medicine could play in transforming Russian schools.

In his analysis of rural schools, Ben Eklof repeatedly emphasized the distinct schooling culture that emerged in post-reform Russia. This culture, he argued, focused on non-coercive motivation, fostering self-esteem and initiative, and differed radically both from the overall Russian realities and from the oppressive classroom that, according to some interpretations, persisted elsewhere in Europe.[37] In his words, the existence of such a "child-centered classroom in a coercive, hierarchical authoritarian society is a major paradox."[38] One important question for me is whether and how school sanitary physicians in Moscow contributed to the construction of this "child-centered classroom," what this approach meant for the material environment of schools and the way children used it.

School sanitary inspection in Moscow began operation in January 1889, and Nikolai Mikhailov, who stood behind this initiative and was ready to do the task for free, was appointed the chief school physician.[39] Already as a zemstvo sanitary physician, Mikhailov advocated for the right of women to practice medicine, particularly at Russian elementary schools. Otherwise, he wrote, "many aspects of the growth and development of the female body, as well as its morbidity and [disease] etiology would for a long time stay in darkness."[40] He maintained this view consistently and hired two female physicians, Olga Andreeva and Olga Gortynskaia, to conduct health inspections in 30 of the 38 girls' schools in Moscow. Those female physicians worked according to the same rules and enjoyed the same salary as their male colleagues.[41]

Mikhailov's reasons behind hiring female physicians included not only women's professional emancipation but also the moral and practical aspects of performing medical control. Considering that, in the 1880s, the system of public health in Moscow was only just emerging, most of the city residents had little contact with (and possibly little trust in) the medical profession. Regular preventive inspection—that is, exposing a child's body, especially a seemingly healthy one, to the medical gaze and intervention— was likely to encounter parental suspicions and resistance. The examination of a female body by male physicians appeared particularly problematic. At the very first meeting of school doctors, Mikhailov suggested:

Girls should not be examined thoroughly, especially by male physicians—at first, it is enough to perform only the examination of the neck, arms, upper chest, head, throat, and the external eye check. Obviously, such examination gives less information than, for example, the examination of the entire skin surface, but considering that the practice of school sanitary inspection is only beginning and that there can be people who do not understand the tasks of the sanitary inspection and misinterpret them, it is better to initially abstain from the thorough examination of girls. If any one of us school men-physicians needs to thoroughly examine a girl, for example, when suspecting syphilis, then probably our comrades, school women-physicians, would not refuse to help us.[42]

Nutrition was another aspect where moral and medical questions conflicted. School doctors observed that a substantial number of pupils at municipal elementary schools suffered from malnutrition. Physicians warned that hunger prevented children from concentrating on their studies and argued that "the organization of proper nutrition should be one of the main and considerable parts of the general hygienic regime of the school."[43]

Yet from the very beginning it became clear that medical and parental ideas of proper child nutrition differed. In line with the dominant scientific ideas on nutrition of the nineteenth century, physicians believed that animal proteins were crucial to a healthy diet and the development of a child. In spring 1889, the Moscow municipal board received several complaints from parents who objected to physicians recommending ferial food—milk, in particular—to children during Lent when Orthodox rules forbade the consumption of any meat, eggs, or dairy products. The head of the municipal school committee, Ivan Lebedev, asked physicians to prescribe ferial food to children only in exceptional, medically justified cases because, as he put it, "one could not go further without disturbing the religious views of the people." The chief school doctor Mikhailov replied to this that physicians could not be deprived of the right to recommend ferial food, especially milk, to undernourished children, if they knew it was necessary for children's health. This could suggest that Mikhailov himself believed physiological laws prevailed over specific rules of religious life. However, physicians agreed that religious views should be respected and that any advice on nutrition should be tentative and careful, "in order not to hurt and insult moral and religious feeling." The final decision on child nutrition was delegated to parents, who were also encouraged to consult priests if they doubted the propriety of milk consumption for their children.[44]

Adequate nutrition at school remained high on the agenda of school doctors for many years. Physicians argued that, according to contemporary hygienic norms, the interval between meals should not exceed four hours, but children were spending between five and seven hours at schools without any provision for meals. The Moscow municipality recognized its responsibility for school lunches and gave a small allowance for these purposes, but with that money the only food schools could provide to their pupils was rye bread. School doctors encouraged parents to give their children home-prepared lunches (in particular, milk), but according to their investigation, about a quarter of all families did not follow that recommendation, and especially during Lent many children ate only bread. Some schools attempted to improve the situation by providing additional free meals (usually milk, meat broth, or porridge) for the weakest and most undernourished children. In 1902, this was reportedly practiced in 45 percent of schools. The entitlement to that additional meal was need-based, and it was school doctors who decided which children would get it.[45]

However, school physicians saw the selective need-based support only as a temporary measure. They insisted that warm lunches should be provided to all pupils at municipal schools, regardless of their social background. According to medical recommendations, those lunches should include milk (at least 300 milliliters per child) and a warm meal—for example, cabbage or potato soup with meat, or rice and millet porridge, or pea soup and buckwheat for the Lenten days. The idea of a universal free warm lunch at schools generally found support in the Moscow municipality, but the continuously increasing number of pupils made it difficult to procure the necessary resources. The eventual solution implied a fifty-fifty participation: in 1911, the municipality decided that schools should offer warm lunches to all pupils, but the meals for the needy half would be financed from the city budget while the wealthier families should cover the expenses from their own means. School lunch was also regarded as a model healthy meal—thus, even if children lived nearby and could go home to have lunch there, this was permitted only if parents could prove that the meal at home was better than the meal at school.[46]

Another major preoccupation of Moscow physicians was the organization of school space. Most municipal schools in Moscow were located in rented premises that were not meant for educational purposes. Opening a municipal school did not imply constructing a specific school building—the scheme that we are used to today. In fact, the link between school as an educational institution and school as a special type of physical space was only emerging: in late imperial Moscow, schools usually occupied only part of a building, sharing it with private apartments. When a proper school building existed, it often housed several legally and educationally independent schools.

Finding school premises was a big problem, partially because of the general shortage of adequate properties, partially for the lack of funds and time. It was school doctors who were responsible for inspecting potential premises and who decided whether those could be converted into schools. In most cases, however, as physicians complained, such a decision required a compromise between hygienic norms and the available properties, and they eventually had to choose "the lesser evil." The fact that rented school premises were all of different quality and design reinforced the role of physicians because no standard solution could be found and a separate evaluation and decision had to be taken in each case. Physicians mobilized their knowledge and resourcefulness to make the available school space more comfortable for children's bodies and more accommodating of their needs. They determined the type of school furniture and its arrangements; they proposed adjustments to the ventilation and heating systems, requested the construction of additional ovens or the reorganization of toilets. Many of these smaller adjustments and recommendations were implemented. That activity, however mundane it might seem, affected the comfort of children at schools and their lived experience of schooling.[47]

Furthermore, constant reports of school physicians on the inadequacy of the rented premises motivated the municipality to construct its own proper school buildings (see figures 10.1 and 10.2). This process developed particularly rapidly in the 1900s, and by 1911, 17 percent of pupils studied in municipal buildings, which were usually shared by several schools. The construction norms for such buildings were developed by architects together with school doctors and reflected many of their previous concerns and recommendations. For example, warm meals for pupils, promoted by physicians, required cooking and eating facilities, and the absence of the latter posed a significant hindrance to the introduction of lunches at school. In their 1904 report on school meals, physicians argued that "a kitchen and a canteen should be recognized as a necessary part of any well-organized school building."[48] Responding to that medical discussion, the new norms required all school projects to include kitchens and canteens. The construction rules also forbade locating any classrooms on the semi-basement floor and established the proper size of rooms and windows. They required separate ventilation and water-based heating systems, with the ability to adjust temperature individually in each room, and stipulated that each building had a room for medical examination. Toilets needed to be heated, with natural light, and equipped with a separate ventilation system and a sufficient number of water closets and sinks with running water (one for every twenty-five students)—a convenience far above the level that most pupils had at home. Moreover, physicians repeatedly emphasized the importance

FIGURE 10.1. Municipal school building on Tsaritsinskaia Street. From *Al'bom zdanii*. National Electronic Library (Russia).

FIGURE 10.2. Municipal school building on Ekaterininskaia Square. From *Al'bom zdanii*. National Electronic Library (Russia).

of games and outdoor activities for schooling, so every school project was required to have recreational rooms and outdoor playgrounds.[49]

Apart from trying to create what they saw as a hygienic and comfortable environment for children's bodies at school, physicians also interacted with children in a more direct way. They measured and weighed them twice a year, organized smallpox vaccinations, and conducted regular medical examinations. From the very beginning, school doctors were discouraged from providing medical care and treatment at schools. Their task was to refer the sick to municipal hospitals and outpatient clinics where they could receive free medical care. In 1903, the city opened a separate municipal outpatient clinic for schoolchildren, which specialized in dentistry and otolaryngology. In 1911, there were already five such outpatient clinics with different specializations serving twelve thousand individual patients annually.[50]

Although theoreticians of school hygiene dreamed of medical experts having control over the entire process of schooling, in the practical work of the Moscow school inspection, issues related to educational processes and curriculum appeared rarely. Moscow school physicians criticized the practice of detaining or delaying pupils after lessons or during breaks and sharply opposed any type of punishment that involved the body. Their general approach was that a child's body needs to be spared. This, however, did not exclude physical exercise.

Physical education became the only matter where medical experts managed to significantly influence the school curriculum. Moscow school doctors advocated physical activity, especially outdoor movement games, as well as the introduction of gymnastics classes not only for boys but also, and especially, for girls. A specific concern was that physical education lessons should never take the form of military gymnastics and drilling, taught by soldiers—as it was practiced at imperial military schools and colleges. Physicians argued that instructors of gymnastics needed to have a background in pedagogy and be trained to work with children. In 1909, the Moscow City Council commissioned a conference on physical education. This conference concluded that physical exercise should be made part of the regular school curricula and that instructors for these classes, as for other school subjects, would need a pedagogical training.[51]

What role did infectious disease and the rise of the germ theory play in the work of Moscow school physicians? In the United States, for example, fighting infectious disease, isolating the sick, and tracing their contacts seemed to be the core of physicians' activity at the turn of the twentieth century and the key reason they were allowed in schools in the first place.[52] The introduction of the school medical inspection in Moscow was also partially motivated by the realization that schools serve as hotbeds of infection and preventing contagion was one of its tasks. School physicians were responsible for reporting cases of infectious disease, introducing quarantine, supervising the isolation of the sick, and approving recovery certificates that would allow children to return to school. In case of an outbreak, school physicians also carried out the disinfection of the school with mercuric chloride or formalin and destroyed the notebooks, textbooks, and drawings of sick pupils. Although these anti-epidemic measures somewhat overlapped with the activity of municipal sanitary physicians, the universal examinations helped identify also less threatening diseases that were beyond the focus of the city sanitary inspection. One important concern was scabies, which, according to Mikhailov's report, was the most common disease among pupils. In general, the work of school physicians played only a limited part in efforts against infectious diseases among schoolchildren, and most cases of such diseases were reported and dealt with by regular physicians, municipal sanitary physicians, and parents.[53]

Crucially, fighting infectious disease never became the dominant function of school physicians in Moscow. Despite the growing acceptance of bacteriology, the public health work at schools—similarly to Russian community medicine in general—never underwent a reductionist turn and was not narrowed down to only interrupting the chain of transmission and fighting germs. In fact, among the questions that led to the strongest mobilization of school physicians—better nutrition, adequate school buildings and furniture, abolition of corporal punishment, sufficient rest, physical exercise, and outdoor games—none was directly linked to infectious disease, but all were centered on environmental and lifestyle changes. These conditions were not seen as a cause of infection but as an aggravating risk factor. In the words of municipal physicians:

> Living in the same conditions as the adults, the young generation certainly is more sensitive to all the deficits of their living environment. In addition to the factors of urban life that negatively affect the health of adults, children also experience the harmful conditions caused by schooling in those unfortunately still common cases when the organization of school life does not fully comply with the requirements of hygiene. This is why epidemic diseases develop among urban schoolchildren so frequently—it is easier for any infection to affect a weakened organism. . . . It is, therefore, necessary to devote all efforts so that schools affect the physical health of pupils to the least possible extent.[54]

In the last imperial decade, school physicians in Moscow received a new responsibility. In 1908, the Moscow municipality opened its first class for "retarded" children. The question of teaching children with special needs, intellectual disabilities, or behavioral problems first appeared on the agenda of Moscow municipal institutions in 1902. The problematization of this question was connected to the rapid expansion of schooling and to discussions on the possible introduction of universal primary education, which meant that even children who had previously been left out of the school system were now brought in contact with it. The practical solution to this question was implemented only in 1908 when the Olginsko-Pyatnitskoe School for Girls opened the first so-called "auxiliary" class. The potential candidates for those classes were identified by teachers or school doctors and underwent a medical and psychological examination by psychiatrists.

The opening of the auxiliary classes signaled a new stage in the medicalization of schooling when psychiatry joined hygiene as an important medical discipline in school life. Even though an alliance between hygiene, psychopathology, and pedagogy is often interpreted as the rise of scientifically rooted segregation and discrimination, it is important to keep in mind the specific context in which such auxiliary classes operated in Moscow. First, in the absence of mandatory schooling, the attendance of both auxiliary and ordinary classes was voluntary. Second, although such classes indeed meant medically defined differentiation between pupils, they were not strictly segregated in the separate institution but existed within ordinary primary schools. Third, because of voluntary attendance and the newness of the initiative, the scope of this project remained small. In 1911–1912, Moscow had sixteen such classes with 252 pupils. Relatively small classes allowed for an individualized approach. The declared goal of those classes was to motivate children to study, to teach them to concentrate and to express themselves, and to give them some basic knowledge about the world. These goals were rooted in the same ideals of the noncoercive classroom and the belief in the importance of physical activity and playing that were advocated by hygienists and reformed pedagogues for the ordinary schools. The key teaching methods included games, drawing, clay modeling, rhythmic gymnastics, special speech exercises as well as long walks, in line with the medical conviction of the value of outdoor activities.[55]

Finally, this new stage did not mean any radical transformation of the role of school doctors or school hygiene. The "auxiliary" classes housed less than 1 percent of all pupils at municipal schools, and although a school physician could indicate a pupil for such classes, the final decision was made by a psychiatrist. In the "ordinary" classes, the cooperation between medicine and pedagogy continued to be expressed primarily not through the language of psychopathology but through that of school hygiene, with its emphasis on social and environmental contexts of health. The 1910 school inspection report provides a good illustration. That year, Moscow had 320 municipal schools with about 53,600 pupils and 15 school physicians. Physicians examined 49,700 pupils and administered 7,800 smallpox vaccinations, disinfected schools 85 times, inspected 89 premises that the municipality planned to rent for schools, and evaluated 24 plans of municipal school buildings. In comparison, only 150 pupils in the entire city had to undergo a special examination that determined whether they should be sent to "auxiliary" classes or could follow the standard curriculum.[56]

Summer Colonies for Weak Children

Despite all the improvement of the school environment and provisions for medical care, neither physicians, municipal reformers, nor educators believed the city to be a healthy environment for children. In 1905, the author of an editorial in *Izvestiia Moskovskoi Gorodskoi Dumy*, the official journal of the Moscow municipality, stated the detrimental impact of living in the city on children's health as an acknowledged fact: "The negative impact of the conditions of urban life on children does not need confirmation; this negative impact particularly affects the children from the poorer classes of the population, who live in highly unhygienic apartments and in a situation of malnutrition. . . . Staying outside of the city, in the countryside environment, revitalizes urban children, strengthens their health, and assists the development of their organism."[57]

Belief in the health dangers of the city environment and the curative powers of the countryside resulted in the project of summer colonies for the pupils of Moscow municipal schools, an initiative that involved both the school administration and the school physicians. Moscow school summer colonies first appeared in 1890. The idea behind this initiative was to allow the weakest and poorest of pupils to spend several months in the countryside in a healthier environment than they could find in Moscow. From the very beginning, such colonies were conceived not as an educational but as a health measure. Although "colonies" (*kolonii*) was the most common naming, these establishments were occasionally referred to as *sanatorii*—a term that would become the standard in Soviet times to mean a type of resort facility that provided both recreational and medical services, usually with full board, for a complex short-term rest, not only for children but also for single adults or families. As one of the municipal physicians put it in 1899: "school summer colonies were a result of the realization that the growing children's bodies of the absolute majority of city pupils have to develop in extremely abnormal conditions and of the desire to do at least something to counterbalance those abnormalities, to give the forming children's bodies the opportunity to develop correctly, even if for a short time."[58]

FIGURE 10.3. Summer colony for children at the municipal estate Lesnoi Gorodok. From *Al'bom zdanii*. National Electronic Library (Russia).

Strictly speaking, Moscow school colonies were not an initiative of the municipality. The idea came from the teachers and the trustees of municipal schools, inspired originally by summer colonies in Switzerland. The plan was financed from school funds so that most children could go there for free. However, the fact that schools had such funds to spend was in itself a result of generous municipal funding that allowed schools to save collected tuition fees for other needs. Teachers and trustees successfully mobilized their social networks to minimize the costs of summer colonies and to get funds to support them. Buildings for summer colonies were always provided free of charge. Usually, these were gentry estates, vacant summer houses, or zemstvo schools in the provinces around Moscow. School money was thus used to cover food, transportation, and service expenses. The municipality became directly involved in the organization of summer colonies only in 1904 and from then on provided an annual subsidy of twelve thousand rubles. This subsidy was financed by a donation of municipal deputy Vasily Bakhrushin that was earmarked to support colonies for poor children. In 1890, there were three summer colonies for 91 children altogether, in 1898 there were twenty-five for 445 children, and in 1911 there were sixty-seven colonies. In 1912 almost 3,000 pupils from Moscow schools spent summer in such colonies (see figure 10.3).[59]

Summer colonies usually lasted for two months. The majority of children came from very poor families, about 40 percent of them were orphans, and about a half of them had never left Moscow before. One common problem in the colonies, which offers a revealing detail to the material status of the Moscow poor, was that many children went on vacation without shoes, and colonies often had to provide footwear on the spot.[60]

These summer colonies were not just a recreational project for children but were seen as an integral part of the school hygiene movement. Even the dictionary entry for "school colonies" in the Brokgauz and Efron encyclopedic dictionary redirected the reader to the section on school hygiene.[61] In line with school hygiene recommendations, colonies emphasized the importance of games and outdoor physical activity but also freedom and creativity. In the first years, some colonies offered educational classes, but this practice was soon stopped, and children spent their time playing (cricket was a particularly popular game), bathing, fishing, gardening, walking in the woods, picnicking, handcrafting, drawing, or organizing amateur theaters and choirs. All those activities were quite normal for the dacha (summerhouse) life of the Russian middle classes, but in the colonies they received a new medical edge—they were not just recreational but healthy and good for child development. Most of these activities were completely new to the pupils of Moscow municipal schools. Educators at the colonies were often shocked by how little children understood of the world outside the city, revealing the assumptions about what a primary school pupil in late nineteenth century Moscow was expected to know. For example, educators complained that many children "could not tell a birch tree from an aspen," "did not know how rye and potatoes grow," and could not distinguish between different types of mushrooms.[62]

A recurrent trope in the colonies' reports echoes the concerns of school hygienists: many urban children did not know how to play, at least in a way that their supervisors recognized as a game and found acceptable. In the words of one of the educators, "especially at the beginning, they [female colonists] were almost unable to play—that is, to invent games by themselves—we always needed to offer them a game, to push, to involve; the only games they knew on their own were 'weddings' and 'mothers and daughters.' To be honest, we very much disliked the latter one: its invariable plot was that the daughter did not obey the mother and did not follow her orders, and the mother scolded and punished her."[63]

In colonies, as in schools, nutrition and malnutrition received particular attention, with a similar emphasis on animal proteins. The board was simple but abundant: milk and bread for breakfast, meat soup and some cereal, potato, or vegetable dish for lunch, tea and bread in the afternoon, usually outdoors, cottage cheese or porridge and milk for dinner. There was no restriction in the size and number of portions, and children were encouraged to eat as much as they wanted (most likely a rare opportunity for many of them). In addition, some colonies had their own vegetable gardens, and colonists also regularly went to pick wild berries (usually strawberries or raspberries) and mushrooms, which were abundant in the forests of European Russia. Although this was a relatively diverse diet, colony organizers were concerned mostly about the consumption of meat and milk, which was monitored, calculated, and served as a measurement of healthy nutrition. According to the 1892 report, an average child consumed about 200 grams of meat and between 600 and 1200 milliliters of milk every day—a dramatic contrast to the diet of the working classes in Central Russia, which was very poor in animal protein.[64]

School doctors were much involved in the project of summer school colonies from the beginning. As the colonies could accommodate only a small number of children, "weakness" became the main selection criteria, and physicians emerged as the judges to identify the "weakest" children most in need of summer vacation in the countryside. They indicated which children should be sent to the colonies, examined and measured them before and after their stay, and visited colonies to check on their health. In the words of Nikolai Mikhailov, "the overall picture of children's health, created by the medical examination on the day before their departure, was most gloomy. They looked more like children collected in hospitals rather than the pupils from municipal schools. In every child one could feel the insufficiency of medicine alone and the need to rely on the mighty powers of nature." The most common health condition that resulted in a referral to a colony was anemia—two-thirds of children in the colonies had this diagnosis in 1891. Considering the emphasis on nutrition, weight gain was vigilantly controlled and interpreted as a sign of the health benefit of such colonies. All children were issued with a special sanitary sheet where their health, weight, height, and chest girth were recorded throughout the stay at the colony. For many health conditions, such as anemia, digestive disorders, or chronic bronchitis, physicians claimed significant improvement during the child's stay in the colony, "even without a pharmaceutical intervention."[65]

Going to summer colonies was voluntary, and parents were not always enthusiastic about sending their children away for two months because many needed their help in the household or at work. Therefore, medically defined "weakness" of children was used as an important argument in negotiations with parents about their children's rights or needs. Physicians maintained that at least 15 percent of all pupils in Moscow were in direct need of such summer vacations for health reasons, and existing colonies could host only a fraction of those. This argument served as a powerful justification for the expansion of the project, giving more and more children the opportunity to spend summer in the countryside, away from the city.[66]

The pupils of municipal schools were the group whose health was in the spotlight of municipal policy. Their numbers were increasing, but because there was no compulsory education, because the average school entry age in Moscow was quite high (about eight years old), and because many left school without completing the short course of study, the school system covered only a minority of the child population of Moscow. While municipal public health efforts focused primarily on those who attended city schools, many school-age children and most children under the age of eight were excluded not only from the school system but also from the sanitary project.

The urban policy toward children and the sanitary project in late imperial Moscow reveal a peculiar combination of consideration and neglect, of medicalization and significant effort to create a healthier childhood and the failure to tackle the most dramatic health problems among children below school age. The question of children's health appeared on the municipal agenda already during the first peak of sanitary reforms in Alekseevan Moscow, and this resulted in the early institutionalization of medical control in the city schools. Contrary to some scholars' observations of the low authority of school doctors, their poor remuneration, their weak influence over governmental decisions, and the sporadic character of their activities, the case of the school medical inspection in Moscow presents a rather different picture.[67] Moscow school doctors had been performing systematic, diverse, and well-paid work since the late 1880s, they were full-time municipal employees and an inherent part of the growing municipal sanitary organization. The authority of school doctors was strong enough not only to transform the school environment and experience of schooling on the micro-level (that is, in each particular school) but also to affect policy on the city level and to prompt changes codified in local norms and regulations.

Importantly, the institutionalization of school medical control in Moscow was motivated primarily by sanitary and socio-environmental concerns. School doctors had already gained significant experience and authority by the time psychiatry started influencing practical medical activity at schools. Although Moscow school doctors were striving for more influence and authority, their declared professional goal was to limit the negative impact of schooling on pupils' health, to construct a comfortable, healthy, and safe environment for poor children, and to compensate for the real and assumed hardships they experienced outside of school.

Medical discourse and expertise were mobilized not to achieve greater discipline but, instead, to promote the ideals of the noncoercive classroom or, rather, to make sure that this classroom was beneficial not only to children's minds but also to their bodies. Similar to their colleagues abroad, Moscow physicians were interested in explaining and categorizing child development, in separating the "normal" from the "abnormal" and the "weak," but it was the physical rather than the mental or moral development that remained their primary concern. Although the introduction of sanitary control in schools subjected children to sometimes rather unpleasant medical manipulations such as regular examinations or smallpox vaccinations, this has to be viewed in the context of an overall voluntary school system, the oppressive and sometimes exploitative treatment of children in their families, and the objective health risks they were facing in Europe's deadliest metropolis. In fact, within the limits of their domain, physicians went perhaps even further than reform-minded pedagogues and advocated freedom, play, rest, and comfort, which in their view could not be sacrificed even for the purposes of education. Indeed, school doctors used medical knowledge to articulate difference—that is, to define "weak" children—but, before the appearance of auxiliary classes and to a substantial degree also afterward, this difference was used not to segregate or to stigmatize but, on the contrary, to support the socially disadvantaged children.

Both the theoretical writings of school hygienists and the practical work of school sanitary physicians enabled looking at education from the perspective of health and environment. They linked the environment of schools, the health of pupils, and children's ability to learn, implying that the municipal government was responsible for the health of those to whom it was providing education. This could perhaps explain why schoolchildren received so much more medical attention compared to other, even more underserved groups of Moscow's child population. School pupils were already a part of the system of municipal public services and, within it, of a particularly successful and symbolically important project—education. As such, they were an identifiable, relatively homogeneous, and spatially concentrated collective that was easy to target and where one could reach out to a large number of people with limited resources.

In Moscow, the school emerged as a place to monitor, promote, and protect the health of urban children and as part of the municipal public health system where children could be examined, diagnosed, given medical advice, measured, weighed, or vaccinated. Even though no treatment was usually performed at school, except for first aid, schools were linked to other municipal medical sanitary services, such as school outpatient clinics or sanitary inspection. The leading part in protecting the health of children was given to publicly employed medical professionals, while parents were assigned only a supporting role. However, the voluntary character of schooling and the short course of study meant that the role of physicians in influencing children's health remained limited and temporary.

Soviet childhood policy borrowed and built upon many of the initiatives that imperial community medicine had introduced or advocated. Medical inspection and centralized vaccination at schools, free territorial outpatient clinics, and summer camps for children are all familiar to researchers of the Soviet period, but it is important to acknowledge that these had all first appeared in the nineteenth century. In fact, many of these initiatives would outlive not only the imperial but also the Soviet regime and would become a part of post-Soviet childhood. In the 1990s, the author of this book spent every summer in an equivalent of a colony for children in the Moscow region (then called a "children's healthifying camp"), and although the institution bore a clear imprint of the Soviet era, some aspects—medical checkups at the beginning and the end of the stay, the importance of weight gain, the emphasis on outdoor activity, and a mandatory glass of milk every day—displayed surprising continuity with summer colonies of the 1890s.

CONCLUSION

Across the Divide

In 1906, Mitrofan Shchepkin—a Moscow journalist, economist, former city deputy, and one of Russia's first urban historians—published a project of a new municipal statute for the city of Moscow. In this project, which would remain unrealized, Shchepkin included a rich commentary with an analysis of the existing government and urban services in Moscow, their role, challenges, and potential development, and summarized his thirty-year experience with the Moscow municipality as both an insider and an expert observer. A part of the project dealt specifically with public health and sanitation—the question, in Shchepkin's view, so big and complex that it permeated and shaped all possible spheres of municipal activity:

> The care for public health and the organization of sanitation touch upon absolutely all aspects of urban life and very different municipal enterprises. Whatever the public government does, starting from paving, cleaning and watering the streets, restructuring the traffic, taking measures to develop trade and industry while protecting the working class and craftsmen, and so on, and ending with the promotion of education in the mass of the urban population and the organization of schooling,—everywhere, in each of these enterprises, the sanitary tasks of the City Public always come up and demand most assiduous attention. Not to mention the largest special municipal enterprises such as

the abattoir, the water pipe and sewage systems—they are fully devoted to the task of public health and called into life by its pressing demands.¹

This quotation reflects a vision of the city that was shared by Moscow municipal leaders, community physicians, and educated elites in the early twentieth century and that Shchepkin himself helped produce. In this vision, the quest for public health appears as a crucial strategy for ordering the city, with sanitation, including sanitary technologies, becoming essential to municipal politics, social policy, and modern urbanism more generally. Yet Shchepkin's evaluation of Moscow's achievement in this sphere is ambiguous. The number of sanitary offices and institutions in Moscow could arouse the envy of any Western European city but, Shchepkin asks, "had the average life of Muscovites improved as a result of these specific measures—who would answer this question?"²

Late imperial Moscow, similar to many other cities in the long nineteenth century, was a site of dynamic reforms in the sphere of public health and sanitation. Those reforms transformed not only urban society but also urban and rural environments and led to the establishment of the new sanitary regime—"the sanitary city"—characterized by medicalization, municipalization, and technologization. Medicalization meant that urban problems were increasingly described and explained in the language of medical science, while a large part of city politics was now centered on medical agendas. Physicians, veterinarians, and medical scientists started to play a significant role in Moscow's urban policy as employees of the municipality and members of numerous committees who advised the municipal government, developed the program of reforms, and put them into practice. Medicalization was a complex multidirectional process that covered not only individual human behavior and conditions but, increasingly, the material environment and collective social processes such as food supply, production, and sale, the disposal and treatment of wastes, the organization of education, the construction of schools and residential buildings that now had to conform with the norms developed by medical experts.

This medicalization was not limited to humans but had an important multispecies dimension. With the rise of veterinary medicine, germ theory, and new understanding of the interconnectedness of human and animal health, livestock animals were incorporated into the sanitary project in Moscow and in Russia more broadly. New medical agendas, including advances in bacteriology, changed how animals were moved, kept, and killed, and their living and dead bodies had to be inspected, described, explained, and treated by human medical professionals. The processes of subordinating human and animal lives to medical control were linked biologically and politically and were inseparable from the development of elected local government systems in the form of zemstvos and municipalities and their quest for greater authority in an autocratic empire.

Municipalization meant that, in late imperial Moscow, the municipality emerged as the main provider of health-care and sanitary services. In 1870, the city owned only one hospital. On the eve of the First World War it had twenty-one municipal hospitals of different specializations, twenty-nine outpatient clinics, ten birthing houses, a sanitary inspection, a trade sanitary inspection, a school sanitary inspection, a veterinary inspection, a sanitary station for laboratory research, a disinfection brigade, a disinfection chamber and laundry, an advanced water supply, a sewerage system with a waste treatment plant, and a public abattoir. Municipal bodies and employees initiated, planned, and delivered health-care and sanitary solutions. In many fields, in the absence of national standards, the municipality was also the source of legal regulation of urban public health. Despite persistent tensions between municipal and imperial authorities, the sphere of health care and sanitation included examples of the imperial administration's strong support for municipal projects and its recognition of municipal expertise and achievement.

What set the Moscow sanitary project apart from the health reforms underway in the zemstvos—and brought it closer to the metropolises in other countries—was its high degree of medical specialization, active international collaboration, and technology-enabled solutions to public health. At the core of zemstvo medicine were generalists whose work comprised not only the treatment of all possible diseases and conditions for the residents of their district but also prophylaxis and hygiene education. This multitasking and the broad definition of medical work were simultaneously results of the lack of resources and an ideological pillar of zemstvo medicine. In Moscow, however, the urban version of *obshchestvennaia meditsina* (municipal public health) was based on the increasing medical specialization that affected the organization of hospitals, outpatient clinics, and sanitary work. Medicine became intrinsically linked with science and technology and so did urban hygiene and sanitation. Sanitary technologies that made it possible to control the safety of water and food, to detect dangerous parasites and pollution, to wash and to disinfect, to supply large volumes of clean water, and to remove and treat wastes produced by households and industries became essential to that version of public health.

The medicalization and technologization of urban life, the municipalization of sanitary services, the rise of medical specialization were all transnational processes; in the late nineteenth century they could be seen in Moscow as they were in many cities across the Western world. These similarities were, of course, not accidental. The image of a sanitary city came to Russia from the West, and sanitary reforms reflected the conscious effort of Moscow elites to make their city not only cleaner and healthier but also more like a European city. More important, perhaps, Russian scientists, physicians, engineers, and urban reformers were active participants in transnational knowledge networks who adapted, combined, and reconfigured innovations from various cities—big and small and not necessarily European—and produced their own.

Forty years of sanitary reforms in Moscow changed the ways in which many Muscovites dealt with disease, what they ate and drank, how they were born, where they lived, worked, and studied, how they interacted with each other and with nonhuman inhabitants of the city. Sanitary reforms contributed to the commodification of animal bodies, broke the association between meat and killing in the eyes of the urban public, and turned slaughter into an abstract and morally unproblematic process, as long as it was hygienic and hidden behind the abattoir walls. The reforms set different standards for both public and private spaces and created completely new types of municipal infrastructures that would shape the environment of the city and its hinterland, which led to a dramatic increase in piped water consumption and reframed waste treatment as a necessary and achievable goal. The reforms triggered debates on industrial pollution; they stimulated the construction of sanitary infrastructure in other cities of the empire and created path dependencies for urban development for decades to come.

The continuing specialization and technologization of urban public health also contributed to a certain crisis of zemstvo medicine in the last imperial decade. After the 1905 Revolution, many zemstvos faced a shortage of personnel who had left the countryside for the cities. The new generation of zemstvo activists and physicians lost faith in the possibility of quick change in the countryside and in the importance of their work. The rise of specialized hospitals and laboratories in the cities challenged the "generalist" model of zemstvo medicine and its ability to address pressing questions of public health. The development of urban sanitary technologies brought into question the adequacy of zemstvo anti-epidemic measures and prophylactic work, which had been one of the ideological cornerstones of zemstvo community medicine. Contemporary critics from the sanitary-technical wing of community physicians, especially Aleksei Sysin, suggested that zemstvos paid too little attention to sanitary infrastructure and that their sanitary work should include not only hygiene education and statistical studies but also engineering solutions such as water supply and sewage systems. This was an important shift as it linked the vision of a healthy Russia to technological rather than cultural progress.[3]

Yet why did those dramatic transformations of the new sanitary city have so little impact on Moscow's mortality statistics? One way to think about it is in terms of equity and access to public health and sanitary services. Planning and building public health infrastructure took time, especially with the limited funds and cumbersome bureaucratic control that were typical of Russian urban government. With the dramatic and unprecedented population growth in Moscow during the post-reform decades, combined with often transient character of migration and the lack of generational continuity, the sanitary city remained fragmented and incomplete. Despite the strong rhetoric of serving the entire urban community behind the sanitary reforms, public services—including hospitals and outpatient clinics, schools and nurseries, sewers and water pipes—were permanently scarce and tended to be concentrated in the inner districts, creating sharp inequalities in use and access. The outskirts of the city could not partake in the new sanitary regime, although by the early twentieth century it was in those underserved districts that population growth and epidemic burden were the highest.

Crucially, however, as I have tried to show in this book, the problem was not only the lack of time and resources to complete the sanitary city. It is important to see the task of making the city "more sanitary" as distinct from the task of making it "less deadly." Although Moscow reformers invested a huge effort in the former, they only occasionally dealt with the latter. The ranking of reforms on the municipal agenda did not follow the logic of "what problem is the major killer." Instead, prioritizing was influenced by various factors: the political aspirations for "Europeanness" and a greater authority and autonomy for the municipality, the ideologies of community medicine, international influences on what a sanitary city should look like and what aspects of urban life should be problematized and changed, the legal and institutional capacity to implement reforms, the power of specific diseases and environmental problems to provoke fear and disgust, along with the personal goals and interests of experts and their ability to convince political leaders. This is why animal health, safe slaughter, and the prevention of zoonotic infections, which had been barely noticeable in Moscow mortality statistics, received much more attention in municipal public health than gastroenteric diseases of children that killed many thousands every year. This is why the health of schoolchildren was guarded by the school sanitary inspection, specialized school outpatient clinics, and summer colonies, while simultaneously tremendous loss of life among infants and young children in Moscow had barely been addressed until the last years of the empire.

On July 19 (August 1), 1914, the Russian Empire entered the First World War. Within weeks, Moscow was flooded with wounded and maimed soldiers. Schools, universities, administrative buildings, cinemas, and restaurants were urgently transformed into makeshift hospitals to accommodate the tens of thousands of soldiers that sanitary trains were bringing to the city from the front.

The war posed an enormous challenge to urban public health. Thousands of municipal employees were drafted into the army. Very quickly Moscow became the most important center of medical evacuation in the country. During the first seventeen months of combat, more than a million wounded and sick soldiers and prisoners of war went through Moscow's distribution centers where they were sorted, registered, washed, had their clothes disinfected, and then were either transported further eastward or sent to Moscow medical institutions. By October 1914, Moscow had more than 500 evacuation hospitals with 38,000 beds, and by 1916, a thousand hospitals with 80,000 beds, the majority of which either belonged to or were maintained by the municipality.

The soldiers were followed by refugees. In the first two years of the war, Moscow received about 450,000 officially registered refugees, and more than 170,000 of them settled in the city. In 1916, the population of Moscow reached two million, an increase of 400,000 compared to 1912, which aggravated the city's already desperate housing situation.[4] Although during the war the city received substantial financial support from the state budget, huge material and administrative resources had to be allocated to distribute, transport, feed, treat, and house soldiers and refugees, to find doctors and nurses, and to procure medical supplies, clothes, and beds.

It would be wrong to see the war years as a crisis for the Moscow sanitary project, however. In fact, in many ways this was the time of its climax. The war gave enormous political influence and freedom of action to community physicians who now grouped themselves around the Union of Zemstvos and the Union of Towns, the newly emerged wartime voluntary organizations. As it was the zemstvos and municipalities that controlled most of the hospitals in the country, their unions quickly turned into the most important medical offices at the rear. The failure of the medical services of the Russian Army endowed community physicians with unprecedented power and funds to direct medical and sanitary work on the home front, and they seized this opportunity with enthusiasm.[5] The Union of Towns was created in August 1914 on the initiative of the Moscow City Council, and the Moscow municipality retained the key role within it throughout the war. The Moscow municipality's experience with sanitary technologies and specialized medical care was now scaled up to the national level as the Union of Towns organized hospitals, outpatient clinics, sanitary trains, bacteriological laboratories, X-ray cabinets, disinfection chambers and laundries, sewage convoys and water filtration stations, at the same time training medical personnel in hygiene, bacteriology, and disinfection techniques.[6]

Remarkably, the new needs and pressures of the war did not result in the discontinuation of peacetime sanitary endeavors in Moscow. Most of them continued and some, such as sanitary services, witnessed a considerable expansion in the very last years of the imperial period. Thanks to the partial launch of the second line of the sewage system, between 1912 and 1916 the number of estates connected to it grew by 1,570 (that is, by 27 percent), and the daily volume of sewage waste it processed jumped from 62,000 to 87,000 cubic meters. Similarly, the average daily consumption of piped water in Moscow increased from 93,000 to 135,000 cubic meters, and 1,700 new estates received access to municipal water pipes.[7] It was in this time that the Moscow municipality started large-scale experiments with sewage treatment through aeration and activated sludge process—the technique that promised to make sewage treatment significantly cheaper and easier to organize (as it required less space) and that would so much impress Charles-Edward Amory Winslow on his visit to Moscow in 1917.[8]

The shortages and the rampant inflation of the war years favored the use of surrogates and food adulteration and gave a new meaning to the market sanitary inspection and the sanitary station, the institutions that had changed little since the time of Alekseev. In 1915, their staff was increased, salaries were raised, and the fight against food fraud became their key responsibility. The sanitary station also carried out systematic bacteriological tests to prevent waterborne and foodborne infections.[9] Municipal schooling continued to expand rapidly during the war years and so did the school sanitary inspection. The number of school physicians increased, and their salaries were raised for the first time since 1889. School summer colonies for weak children also continued operation throughout the war, with 2,570 Moscow children spending the summer in such colonies in the countryside in 1916. Even the public health of infants, that stepchild among the sanitary initiatives of the Moscow municipality, saw some expansion in that period, with the creation of additional milk kitchens and the opening of a new department for infants at the Morozov Hospital, the one that Winslow admired in 1917.[10]

Of all the institutions discussed in this book, only the work of the municipal abattoir was radically disrupted by the war. The overload of the transport system, the ban on cattle transportation by railway, and the mass culling of livestock in the country during the war meant that supply to the abattoir dropped dramatically and meat in Moscow became a rarity. The area of the abattoir that used to house thousands of animals on their way to slaughter became the site of Moscow's largest refugee camp, which now served as home to 4,500 displaced humans who fled the slaughter of the war.[11]

It is important to consider these last years of the imperial period when evaluating the Moscow sanitary project, especially because Moscow urban planners and municipal community physicians were obviously unaware that their deadline would be set for 1917. Despite the huge influx of wounded soldiers and refugees, the overcrowding, and the epidemics in the western and southern provinces, the epidemiological situation in Moscow remained surprisingly stable throughout the early war years. There were outbreaks of cholera, typhus, and dysentery, but their death toll remained moderate, not least because of the efforts of the Moscow medical authorities and the resilience of the public health and sanitary system.[12] Simultaneously, the war caused a sharp decline in natality and a corresponding decline in infant deaths. In 1916, the general mortality rate in Moscow fell to 23 per 10,000, the lowest of the entire imperial period.[13] From the point of view of demographic statistics, Russia's sanitary city had never looked so good as during the war, in the very last year of the empire.

In March 1917, the tsar Nicholas II abdicated, and the empire gave way to a republic headed by the Provisional Government. Although this was such a turbulent and decisive time in Russia's political history, it seems a strangely unremarkable year for Moscow's public health, as mortality remained relatively low and medical and sanitary structures showed surprising endurance and continuity with previous years. In November 1917, the Bolsheviks took power in Moscow. In December, the Moscow City Council held its last meeting. In 1918, Moscow became the capital of the new Soviet state. In the following years the sanitary city collapsed.

While the Bolshevik leaders and many of the former community physicians were busy creating a new system of socialist medicine, the acute fuel crisis and the lack of maintenance during the revolutionary chaos (especially after the radical housing reform of 1918 that municipalized apartment buildings) led to freezing of plumbing pipes in the cold winter months and massive damage to the sewage and water supply systems in Moscow. Residents of apartment houses used their backyards as improvised privies and, especially when the snow began to melt, Moscow was literally flooded with excrement. Diaries from 1919 and 1920 provide vivid descriptions of Moscow's filth and devastation: broken pipes and water closets, overflowing sewage, streets and yards covered in rotting waste and animal carcasses.[14] "The dirt is unmanageable, the stench horrible," wrote Nikita Okunev, an employee of a Moscow steamboat company, who lived in such a building. "This is Beijing and not Moscow, or the Moscow of one hundred years ago."[15]

The collapse of the sanitary city combined with dire food shortages and a deficit of medications and disinfectants led to a tremendous crisis of public health and loss of life.[16] Mortality rates in Moscow skyrocketed, reaching the astonishing number of 454 per 10,000 in 1919. Many Muscovites left the city, fleeing political persecution, violence, hunger, cold, filth, and epidemics—its population decreased by half a million compared to the eve of the revolution. Despite the rapid population decline, 20,000 more people died in Moscow in 1919 than in 1916. The major killer of that year was typhus, the disease that thrived in conditions when bathing, heating water, changing and washing clothes became impossible. In 1919, more than 12,000 people perished of typhus, compared to less than 200 in 1916. If 1916 was Moscow's best year in terms of mortality statistics, 1919 was the deadliest since the beginning of statistical recording.[17]

Crises are known to be great revealers, and the violent and chaotic early postrevolutionary years in Moscow exposed the quiet but dramatic transformation that had taken place in the last decades of the imperial period and the new vulnerabilities that it had generated. Within a short time, sanitary services and technologies started to be perceived as indispensable for the city and were embedded in the urban life to such an extent that their malfunction produced a devastation comparable to a military invasion. It is remarkable how Okunev's description echoes the debates of the nineteenth century. If, for municipal reformers of the 1870s and 1880s, urban sanitation and the sewage system symbolized progress and Europeanness, in 1920 the breakdown of sanitary services meant a transition from Europe to Asia, a return to a different century, a sign of a civilizational collapse.

Because of its—in fact, very recent—indispensability and embeddedness in urban life, the sanitary project of imperial Moscow was quickly reassembled into the new Soviet public health. This process was helped by some remarkable personnel continuities, both in specific municipal organizations and at the highest level, as the sanitary epidemiological work of the People's Commissariat of Health was now led by Aleksei Sysin, a former Moscow sanitary doctor and a mouthpiece of a sanitary-technical wing of Russian community physicians. This was both a material and an ideological fit. Moscow sanitary services were not only already conveniently there (although in urgent need of reconstruction) for the new authorities to claim and reuse them. Many goals and features of the imperial sanitary project—modernization, medicalization, centralization, technologization, inspectability and control, subordinating the unruly nature and the needs of rural communities to the demands of the industrial city and the rhetoric of collective good—matched the ideological imperatives of the new power.

The scale and novelty of sanitary reforms in late imperial Moscow set it apart from other Russian cities and ensured that its position as a sanitary city was quite unique and its experience, while important for understanding late imperial modernity, was hardly representative of broader urban Russia, where municipal public health reforms started only in the 1900s and 1910s.[18] Although the new power promised to overcome the inequalities of the empire and to bring health and prosperity to workers and peasants across the country, public health was quickly sacrificed to the goals of industrialization and economic production, and Moscow continued to remain a rare sanitary city in Soviet Russia. Large-scale investments in sanitary reforms and technologies in other cities and towns were postponed until the postwar decades, while most of the Russian countryside still has not seen the development of sanitary services that Sysin had advocated for back in the 1900s. Even at the turn of the 1950s, in terms of access to sanitary services, Moscow was, as Donald Filtzer puts it, "far and away the most advanced city in the country."[19] Was it just because of its privileges as a capital and a model Soviet city? Or was it a long-term effect of the reforms conceived in the 1880s?

NOTES

Introduction

An earlier version of chapter 3 was published as Anna Mazanik, "Industrial Waste, River Pollution and Water Politics in Central Russia, 1880–1917," *Water History* 10 (2018): 207–22, reproduced with permission by Springer Nature. An earlier version of chapter 5 appeared in Anna Mazanik, "'Shiny Shoes' for the City: The Public Abattoir and the Reform of Meat Supply in Imperial Moscow," *Urban History*, 45, no. 2 (2018): 214–32, © Cambridge University Press, reproduced with permission. Selected passages in chapters 1 and 2, as well as sections of chapter 3 have been adapted from Anna Mazanik, "Learning from Smaller Cities: Moscow in the International Urban Networks, 1870–1910," in *Interurban Knowledge Exchange in Southern and Eastern Europe, 1870–1950*, ed. Eszter Gantner, Heidi Hein-Kircher, and Oliver Hochadel (New York: Routledge, 2021), 119–33, reproduced with permission by Taylor and Francis Group LLC (Books) US through PLSclear. Sections of chapter 6 were originally published in Mazanik, "Public Health across Species." Parts of chapter 10 appeared as Mazanik, "School Doctors," in the yearbook of the New Europe College, in fulfillment of requirements of the Pontica Magna Fellowship.

1. Borrero, *Hungry Moscow*, 11–20. The complex impact of the war and revolution on life in the cities across the Russian Empire has recently been studied in Miller and Chernyi, *Goroda imperii*.
2. Winslow, "Public Health Administration," 2219.
3. Winslow, "Public Health Administration," 2216–17.
4. Winslow, "Public Health Administration," 2200, 2216. Winslow included a description of those sewage experiments in Moscow in the second edition of the study of sewage disposal he coauthored. See Kinnicutt, Winslow, and Pratt, *Sewage Disposal*.
5. On Soviet narratives of the medicine in the Russian Empire, see Zatravkin and Vishlenkova, *"Kluby" i "getto"*, 19–42.
6. Bradley, *Muzhik and Muscovite*, 324–33; Thurston, *Liberal City*, 8, 19–20;

Colton, *Moscow*, 56–57; Ruble, *Second Metropolis*, 276–86. Similar accounts of Russia's extreme backwardness in terms of public health can be also found in works on other cities. See, for example, Smith, "Public Works in an Autocratic State."

7. Filtzer, *Hazards of Urban Life*, 7–10.

8. Frieden, *Russian Physicians*; Hutchinson, *Politics and Public Health*; Solomon and Hutchinson, *Health and Society*; Hachten, "Science in the Service of Society"; Beer, *Renovating Russia*; Becker, *Medicine, Law, and the State*.

9. Evans, *Death in Hamburg*; Hardy, *Epidemic Streets*; Snowden, *Naples in the Time of Cholera*; Hamlin, *Public Health and Social Justice*; Hietala, *Services and Urbanization*; Melosi, *Sanitary City* (2000); Melosi, *Garbage in the Cities*; Barnes, *Great Stink of Paris*; Echenberg, *Plague Ports*; Mosley, *Chimney of the World*. On animals in the city, see Cronon, *Nature's Metropolis*; Biehl, *Pests in the City*; Robichaud, *Animal City*.

10. Gleason, "Public Health, Politics and Cities"; Walker, "Public Health"; Henze, *Disease, Health Care and Government*; Davis, *Russia in the Time of Cholera*. See also chapter 1 in Williams, *Health and Welfare*.

11. Martin, "Sewage and the City"; Agafonova, "Urban Pollution"; Agafonova, "Water Supply to the Small Cities"; Vinogradov, "Kazan' Citizens against Air Pollution."

12. Moon, *The Plough That Broke the Steppe*; Jones, *Empire of Extinction*; Obertreis, *Imperial Desert Dreams*; Breyfogle, *Eurasian Environments*; Peterson, *Pipe Dreams*; Samojlik, Fedotova, Daszkiewicz, and Rotherham, *Białowieża Primeval Forest*. There has been considerable effort made to study water environments in Saint Petersburg. See, for example, Dills, "River Neva"; Kraikovski and Lajus, "Living on the River."

13. See, for example, Egorysheva, Sherstneva, and Goncharova, *Meditsina gorodskhikh obshestvennykh samoupravlenii*; Sherstneva, "Moskovskoe gorodskoe samoupravlenie"; Denisov, *Stranitsy istorii sanitarnogo dela*.

14. An important exception here is Pirogovskaia, *Miasmy, simptomy, uliki*, although it has a specific focus on smell.

15. Elina, *Ot tsarskikh sadov*; Ozerova, "Istoriia izucheniia"; Loskutova and Fedotova, *Stanovlenie prikladnykh biologicheskikh issledovanii*; Davydov, "Vodosnabzheniie."

16. Saunier and Ewan, *Another Global City*; Gantner, Hein-Kircher, and Hochadel, *Interurban Knowledge Exchange*.

17. Worboys, *Spreading Germs*; Worboys, "Was There a Bacteriological Revolution?"; Barnes, *Great Stink of Paris*.

18. A rare exception is the study of the controversial experience of the Odessa Bacteriological Station in the unpublished dissertation by Elizabeth Hachten. See Hachten, "Science in the Service of Society"; some of its findings are summarized in Hachten, "In Service of Science and Society." Lisa Walker discusses the practical

applications of bacteriology in the early twentieth century in her unpublished dissertation. See Walker, "Public Health," 154–97.

Chapter 1. Discovering Moscow's Dirt

1. "Doklad N 55," 22. Throughout this book, unless otherwise noted, all translations from the original Russian are my own.

2. *Budil'nik* 38 (1884): 455.

3. Martin, "Sewage and the City."

4. Verner, *Sovremennoe khoziaistvo*, 6; *Statisticheskii ezhegodnik g. Moskvy*, 7; *Glavneishie predvaritelnye (1912)*; Zviagintsev, *Moskva*, 112; Mazanik, "City as a Transient Home."

5. Anderson, *Internal Migration*, 106; Mironov, *Social History*, 333, 341–42.

6. Bradley, *Muzhik and Muscovite*, 103; Mazanik, "City as a Transient Home," 54–56.

7. I discuss the relations between gender and migration in Moscow in Mazanik, "City of Men." See also Mironov, *Social History*, 347; Burds, *Peasant Dreams*, 25–38.

8. "Novoe Vremia" newspaper cited in Buryshkin, *Moskva kupecheskaia*, 98.

9. On the Moscow business elite, see Owen, *Capitalism and Politics*; Rieber, *Merchants and Entrepreneurs*; Petrov, *Moskovskaia burzhuaziia*; Ruble, *Second Metropolis*, 80–87.

10. Todes, *Ivan Pavlov*, 22–23; Frieden, *Russian Physicians*, 53–54; Byford, *Science of the Child*, 27–28.

11. Nardova, *Samoderzhavie i gorodskie dumy*; Nardova, *Gorodskoe samoupravlenie*, 82–108; Pisar'kova, *Gorodskie reformy*, 313–31.

12. Pisar'kova, *Gorodskie reformy*, 129–30.

13. Frieden, *Russian Physicians*, 47–52, 93–94; Todes, *Ivan Pavlov*, 22–23.

14. Hachten, "Science in the Service of Society," 252–53; *Istoriia Imperatorskoi Voenno-meditsinskoi Akademii*, 554–55.

15. Bonner, "Rendezvous in Zurich," 10–12.

16. Hachten, "Science in the Service of Society," 254; Boubnoff, *Institut d'hygiène*, 3; Mikhailov, *Pamiati professora F. F. Erismana*, 5. The works of the hygienic laboratory and the Institute of Hygiene were published as Erisman, *Sbornik rabot gigienicheskoi laboratorii*.

17. For example, in 1887, the deputies of the city council in Nizhnii Novgorod asked Erismann to recommend a candidate to fill the new post of municipal sanitary physician. On his recommendation, the post went to Pavel Rozanov. See Walker, "Public Health," 36.

18. Hachten, "Science in the Service of Society," 35.

19. Erisman, *Kurs gigieny* (1887), 1:9, 7.

20. Erisman, *Kurs gigieny* (1887), 1:10 (also 7–10).

21. Erisman, "Sanitariia," 261.

22. Hamlin, *Public Health and Social Justice*, 52–83; La Berge, *Mission and Method*, 82–100; Melosi, *Sanitary City* (2008), 28–68.

23. Worboys, *Spreading Germs*; Worboys, "Was There a Bacteriological Revolution?"; Barnes, *Great Stink of Paris*.

24. Hachten, "Science in the Service of Society"; Hutchinson, "Tsarist Russia"; Walker, "Pen and the Test-Tube"; Davis, *Russia in the Time of Cholera*, 62–67.

25. Hutchinson, *Politics and Public Health*, 35–36; also Henze, *Disease, Health Care and Government*, 23. For a more nuanced analysis of Erisman's relation to contagionism and bacteriology in the context of debates on cholera, see Davis, *Russia in the Time of Cholera*, 58–62.

26. Hachten, "Science in the Service of Society," 266–83; Hutchinson, "Tsarist Russia"; Krug, "Debate over the Delivery of Health Care," 239–40.

27. Erisman, "Znachenie bakteriologii dlia sovremennoi gigieny," 20 (also 19–20, 25–27).

28. Erisman, *Novye kliniki i instituty*, 134–38.

29. Erisman, *Kurs gigieny* (1887), 1:102, 115–18, 245–47.

30. Erisman, *Kurs gigieny* (1892), 1:122–34, 139.

31. Barnes, *Great Stink of Paris*, 3.

32. Frede, *Doubt, Atheism*, 13–16, 218–19; Knight, "Was the Intelligentsia Part of the Nation?"

33. Frieden, *Russian Physicians*, 313–14; Walker, "Public Health."

34. Erisman, "Organizatsiia obshchestvennoi gigieny v Rossii," 226.

35. Erisman, "Organizatsiia obshchestvennoi gigieny v Rossii," 251.

36. Alexander, *Bubonic Plague*, 38; Frieden, *Russian Physicians*, 21–52.

37. "Ustav vrachebnyi," in *Svod zakonov Rossiiskoi Imperii* (Saint Petersburg: 1892), 1–2; *Polnyi Svod Zakonov Rossiiskoi Imperii*, coll. II, vol. 17, part 1 (Saint Petersburg, 1843), Nr. 15202. See also *Polnyi Svod Zakonov Rossiiskoi Imperii*, coll. III, vol. 241 (Saint Petersburg, 1907), no. 24254; Freiberg, "Vrachebno-sanitarnoe zakonodatelstvo," 182–85; Hutchinson, *Politics and Public Health*, 4–6; Davydov, *Moskva, vek XX*, 1:8; Velychenko, "Chislennost' biurokratii i armii." For the study of railway medicine, which illustrates the compartmentalization and complexity of the Russian public health system, see Strobel, *Die "Gesundung Russlands."*

38. Elisa Becker has argued, however, that the medical profession remained strongly dependent on the state also after the Great Reforms and that physicians strove to redefine their role from within the state rather than from outside it. See Becker, *Medicine, Law, and the State*, particularly p. 270.

39. "Dnevnik Obshchestva vrachei Kazani," 3 (1873), cited in Strashun, *Russkaia obshchestvennaia meditsina*, 11.

40. Zhbankov, "Kratkie svedeniia," 39.

41. Frieden, *Russian Physicians*; Strobel, *Die "Gesundung Russlands"*; Kuz'min,

Vlast', obshchestvo i zemskaia meditsina; Koroleva, Karelin, and Pisar'kova, *Zemskoe samoupravlenie v Rossii*; Tretiak, *Stanovlenie i razvitie zemskoi meditsiny i veterinarii*; Moiseieva, *Zemskaia meditsina Simbirskoi gubernii*. There are also a number of unpublished dissertations in Russian on regional aspects of zemstvo medicine.

42. See, for example, Henze, *Disease, Health Care, and Government*, 46.

43. Krug, "Debate over the Delivery of Health Care," 231–32.

44. Erisman coordinated this research and authored some of the volumes. See Erisman, *Sanitarnoe issledovanie fabrichnykh zavedenii Moskovskogo uiezda*; Erisman, *Sanitarnoe issledovanie fabrichnykh zavedenii Klinskogo uiezda*.

45. Gantner, Hein-Kircher, and Hochadel, *Interurban Knowledge Exchange*, especially Gantner, Hein-Kircher, and Hochadel, "Introduction"; Hietala, *Services and Urbanization*; Saunier and Ewan, *Another Global City*; Behrends and Kohlrausch, *Races to Modernity*.

46. Hard and Misa, "Modernizing European Cities"; Gantner, Hein-Kircher, and Hochadel, "Introduction," 6–17.

47. Hutchinson, *Politics and Public Health*, 69–72.

48. See also "Zaiavlenie glasnogo A.D. Lopasheva," 1–2; "Doklad N 55," 1, 22; V. F[idler], *Moskva*, 89–90.

49. Martin, "Sewage and the City," 268–73.

Chapter 2. Making a Sanitary City

1. *Rus* 10 (1881): 4.

2. Poleshchuk, "Deiatel'nost' N. A. Alekseeva," 33–42.

3. Vishnevskii, *Kniaz' Vladimir Andreyevich Dolgorukov*, 39–40.

4. *K. P. Pobedonostsev i ego korrespondenty*, 268.

5. See letters of V. D. Dolgorukov in *K. P. Pobedonostsev i ego korrespondenty*, 346 (also 266–68, 344–46).

6. On the conflicts between the mayor and the governor-general, see Poleshchuk, "Nevozmozhnaia v Moskve dolzhnost'," 106–7 (107).

7. These words are mentioned in the diary of A. S. Suvorin, quoted in Poleshchuk, "Nevozmozhnaia v Moskve dolzhnost'," 108.

8. Bibikova, "Politicheskaia politsiia," 516–19.

9. Brower, *Russian City*, 95 (also 91–139).

10. The term "Alekseevan regime" was used by the historian Grigorii Dzhanshiev in *Epokha velikikh reform*.

11. Chicherin, *Vospominaniia*, 360, 429.

12. Pisar'kova, *Gorodskie reformy*, 274–75.

13. Kotsonis, *Making Peasants Backward*, 1–8.

14. *Izvestiia Moskovskoi Gorodskoi Dumy* 7 (1885): 787–88.

15. Rieber, *Merchants and Entrepreneurs*, 139–48, 165–77.

16. See, for instance, discussions of sanitary reforms in *Izvestiia Moskovskoi*

Gorodskoi Dumy 7 (1885): 779–88; Tsentral'nyi gosudarstvennyi arkhiv goroda Moskvy (Central State Archive of the City of Moscow; hereafter cited as TsGA Moskvy), 179:58:44:17–18. Throughout, in the archival references, the first number refers to *fond*, the second to *opis'*, the third to *delo*, and the fourth to the pages.

17. Morris, "Governance."
18. *Izvestiia Moskovskoi Gorodskoi Dumy* 10 (1885): 1113.
19. TsGA Moskvy, 179:58:30:13–16, 63–64; *Izvestiia Moskovskoi Godoskoi Dumy* 7 (1885): 787.
20. *Sbornik obiazatel'nykh dlia zhitelei g. Moskvy*, 24–41; *Sbornik obiazatel'nykh dlia zhitelei goroda Moskvy*, 1–131.
21. Zhbankov, *Sbornik po gorodskomu vrachebno-sanitarnomu delu*, 27–28.
22. *Izvestiia Moskovskoi Gorodskoi Dumy* 4 (1887), appendix, 8; *Izvestiia Moskovskoi Gorodskoi Dumy* 8 (1887), section 1, 10; Chertov, *Gorodskaia meditsina*, 86–89.
23. "Kratkii otchet o deiatel'nosti gorodskikh ambulatorii"; Uspenskii, *Moskva*, 42–46; Pechorkin, "Ambulatoriia."
24. *Finansy goroda Moskvy*, 61–68.
25. *Otchety moskovskikh gorodskikh sanitarnykh vrachei*.
26. Uspenskii, *Moskva*, 22–23.
27. On prostitution and syphilis in imperial Russia, see Engelstein, *Keys to Happiness*; Bernstein, *Sonia's Daughters*; Hearne, *Policing Prostitution*.
28. I discuss this reform in detail in my dissertation, Mazanik, "Sanitation," 109–12.
29. TsGA Moskvy, 179:58:30:64.
30. TsGA Moskvy, 179:58:44:8–20.
31. Chertov, *Gorodskaia meditsina*, 100–110; Zhbankov, *Sbornik po gorodskomu vrachebno-sanitarnomu delu*, 27–49.
32. *Vodosnabzhenie goroda Moskvy*, 27–62.
33. On the growing importance of the public domain and public property in post-Reform Russia, see Pravilova, *Public Empire*.
34. Pisar'kova, *Gorodskie reformy*, 190, 212; Nardova, *Gorodskoe samoupravlenie*, 61–66; Nardova, *Samoderzhavie i gorodskie dumy*, 12–13, 29; Brower, *Russian City*, 123.
35. See Pisar'kova, *Gorodskie reformy*, 212–20.
36. Owen, *Capitalism and Politics*, 160–61.
37. Ratmanov, *Sovetskoe zdravookhranenie*, 285.
38. Vitte, *Vospominaniia*, 2:348.
39. Vishniakov quoted in Pisar'kova, *Gorodskie reformy*, 400.
40. See, for example, Vitte, *Vospominaniia*, 2:350. The comments on the army-like behavior of Trepov were given not only by his enemies but also by his allies, such as, for example, the director of the Moscow security police Sergei Zubatov,

see the letter of Zubatov to V. L. Burtsev in Koz'min, *Zubatov i ego korrespondenty*, 70–71.

41. Henze, *Disease, Health Care, and Government*, 51. The cholera epidemic in Hamburg and Italy is discussed in Evans, *Death in Hamburg*; Snowden, *Naples in the Time of Cholera*.

42. Frieden, "Russian Cholera Epidemic"; Frieden, *Russian Physicians*, 143–60, 325; Henze, *Disease, Health Care, and Government*, 3, 51; Hutchinson, *Politics and Public Health*; Figes, *People's Tragedy*.

43. TsGA Moskvy, 16:130:240:25, 34–35. I discuss the political response to the Moscow cholera outbreak in Mazanik, "Managing the Catastrophe."

44. TsGA Moskvy, 16:130:240:23–35; *Izvestiia Moskovskoi Gorodskoi Dumy* 7–8 (1892), 179.

45. *Polnoe sobranie zakonov Rossiiskoy Imperii*, 3rd ed. (Saint Petersburg: Gosudarstvennaia Tipografiia, 1885), vol. 1, no. 350; Daly, "On the Significance of Emergency Legislation," 609, 617.

46. TsGA Moskvy, 16:130:240:23–25.

47. TsGA Moskvy, 16:130:240:43, 46, 49, 53, 63.

48. TsGA Moskvy: 16:130:240:130–31.

49. Davydov, *Moskva, vek XX*, 37.

50. Mazanik, "Managing the Catastrophe."

51. For the discussion of the 1905 Revolution in Moscow, see Engelstein, *Moscow, 1905*.

52. Pisar'kova, *Gorodskie reformy*, 234–38.

53. Verner, *Sovremennoe khoziaistvo*, 108–13.

54. Uspenskii, *Moskva*, 32–48.

55. Verner, *Sovremennoe khoziaistvo*, 153–55, 281, 389; *Moskovskaia gorodskaia duma*, 130.

Chapter 3. The Sewage System

1. Goubert, *Conquest of Water*; Melosi, *Sanitary City* (2000); Winiwarter, "Where Did All the Waters Go?". See also Benidickson, *Culture of Flushing*; Otter, "Locating Materiality"; Schneider, *Hybrid Nature*.

2. Melosi, *Sanitary City* (2000), 93; "Doklad N 55," 4–12.

3. *Statisticheskii ezhegodnik Rossii za 1915 g.*, section 5, 4–5. However, as definitions of what constitutes a sewage system varied at the time, there could be alternative calculations.

4. Late imperial debates on the construction of a sewage system in Saint Petersburg are analyzed in Malinova-Tsiafeta, *Iz goroda na dachu*, 120–56.

5. Barnes, *Great Stink of Paris* 3; Worboys, "Was There a Bacteriological Revolution?"

6. *Poiasnitel'naia zapiska*, 1; Gerasimov, *Istoriia uluchsheniia Moskvy-reki*, 4.

7. Kastal'skii, *Moskva v sovremennom sostoianii*, 9.

8. Petunnikov, "Sostav i svoistva Moskovskikh vod," 15; "Otchet sanitarnogo vracha Prechistenskoi chasti"; "Sanitarnyi otchet po Iauzskoi chasti"; "Otchet sanitarnogo vracha Arbatskoi chasti"; "Otchet sanitarnogo vracha Meshchanckoi chasti"; Zimin, "K voprosu o vodosnabzhenii Moskvy."

9. Astrakov, "O kolichestve vody, protekaiushchei v reke Moskve."

10. "Doklad N 55," 22, 37.

11. Ostroglazov, *Smertnost' v Moskve* (1887), 1; *Trudy Moskovskogo gorodskogo statisticheskogo otdela*, 2:68–69.

12. Popov, "Kanalizatsiia goroda Moskvy," 6.

13. *Sbornik obiazatel'nykh dlia zhitelei g. Moskvy i chastiiu drugikh gorodov postanovlenii*, 30–31.

14. TsGA Moskvy, 16:130:240:1, 7–8; *Sbornik obiazatel'nykh dlia zhitelei g. Moskvy i chastiiu drugikh gorodov postanovlenii*, 31, 53; "Otchet sanitarnogo vracha Prechistenskoi chasti," 44–55; "Sanitarnyi otchet po Iauzskoy chasti," 34–44; "Otchet sanitarnogo vracha Arbatskoy chasti"; "Otchet sanitarnogo vracha Meshchanckoi chasti"; Kastal'skii, *Moskva v sovremennov sostoianii*, 8, 14, 18–19; Kamenetskaia, *Kanalizatsiia goroda Moskvy*, 1; Nikitin, "Kanalizatsiia," 287.

15. TsGA Moskvy, 179:33:4:4.

16. TsGA Moskvy, 179:33:4:4.

17. TsGA Moskvy, 179:33:4:11. A similar account was given by Kastal'skii, *Moskva v sovremennom sostoianii*, 19.

18. Petunnikov, "Sostav i svoistva Moskovskikh vod"; Petunnikov, "Materialy dlia izucheniia Moskvy"; Petunnikov, "Gidrografischeskiy ocherk Moskvy"; Kotsyn, *Opyt sistematicheskikh nabliudenii*, 157–63 (esp. 161–62). The history of the scientific study of the Moskva River is discussed in great detail in Ozerova, "Istoriia izucheniia gidrograficheskoi seti basseina r. Moskvy."

19. Erisman, *Kurs gigieny* (1887), 1:192, 199.

20. Popov, *Domovye i dvorovye stoki*, 1; Popov, "Kanalizatsiia goroda Moskvy," 26–32; "Sanitarnyi otchet po Iauzskoy chasti," 35; "Otchet sanitarnogo vracha Prechistenskoy chasti," 46; "Doklad N 55," 3; TsGA Moskvy, 179:33:4:4.

21. Sokolov, *Rezultaty analizov vody reki Iauzy*, 10–11 (also 8–10).

22. Pirogovskaia, *Miasmy, simptomy, uliki*, 391.

23. "Zaiavlenie glasnogo A.D. Lopasheva"; "Doklad N 55," 1, 22; Kotsyn, *Opyt sistematicheskikh nabliudenii*, 161–62; V. F[idler], *Moskva*. 89–90.

24. Petunnikov, "Moskva i eia budushchnost'," 9.

25. Frieden, *Russian Physicians*, 325; Henze, *Disease, Health Care, and Government*, 11.

26. *Smertnost' naseleniia goroda Moskvy*, 85–86.

27. Popov, "Kanalizatsiia goroda Moskvy."

28. *Izvestiia Moskovskoi Gorodskoi Dumy* 4 (1880), 15–20.

29. Nikitin, "Kanalizatsiia," 288–89.
30. Nikitin, "Kanalizatsiia," 290.
31. "Obzor podgotovitelnykh rabot," xii–xiii.
32. *Izvestiia Moskovskoi Gorodskoi Dumy* 9 (1881), 993.
33. *Trudy III (Stroitel'nogo) otdela*, 51–68.
34. Melosi, *Precious Commodity*, 38.
35. *Trudy III (Stroitel'nogo) otdela*, 131–53, 193–231, 273–80; Fedorov, *Po povodu trudov Komissii*, 4–15.
36. Kastal'skii, *O razdel'noi sisteme*.
37. Butler and Davies, *Urban Drainage*, 20–26.
38. Kastal'skii, *O razdel'noi sisteme*.
39. Tarr, "Separate vs. Combined Sewer Problem," 313–14.
40. Tarr, "Separate vs. Combined Sewer Problem," 316–19; Melosi, *Sanitary City* (2000), 153–59.
41. Verner, *Sovremennoe khoziaistvo*, 4.
42. On the difference between combined sewers in Western Europe and in the United States, see Melosi, *Sanitary City* (2000), 93.
43. Melosi, *Sanitary City* (2000), 155.
44. Kastal'skii, *O razdel'noi sisteme*, 15.
45. The protocols of the meeting are published as an appendix to Kastal'skii, *O razdel'noi sisteme*, 33–106 (Erismann's position is discussed on pp. 68–69).
46. Kastal'skii, *O razdel'noi sisteme*, 48–49.
47. Kastal'skii, *O razdel'noi sisteme*, 83–89.
48. Kastal'skii, *O razdel'noi sisteme*, 106.
49. *Izvestiia Moskovskoi Gorodskoi Dumy* 10 (1887), 864, 867–68.
50. This argument has been elaborated, for example, in Evans, *Death in Hamburg*.
51. *Izvestiia Moskovskoi Gorodskoi Dumy* 10 (1887), 860.
52. There was, however, another sewerage project, submitted to the municipality by N. Grunner. This project proposed district-level systems that would collect wastes in big reservoirs; it was quickly rejected.
53. *Zhurnaly Komissii* (1894): 50, 139–44.
54. Hard and Misa, "Modernizing European Cities," 6–8; Arnold, *Science, Technology and Medicine*, 1–18.
55. Zviaginskii, *Domovaia kanalizatsiia*, 11.
56. Frenkel, *Osnovy obshchego gorodskogo blagoustroistva*, 231. See also Kastal'skii and Kozhinov, *Vodosnabzhenie i kanalizatsiia*, 8, 16.

Chapter 4. What Happened to Waste?

1. *Pervaia Vseobshchaia perepis' naseleniia*, 2:62–63. The data refers to mother tongue statistics.

2. *Izvestiia Moskovskoi Gorodskoi Dumy* 10 (1887): 870 (also 868–74).

3. *Glavneishie predvaritelnye dannye perepisi g. Moskvy 31 ianvaria 1902 g.*, 1:32; Verner, *Sovremennoe khoziaistvo*, 4.

4. Kleptsov, "O syrosti kamennykh zdanii"; Bogopol'skii, "K voprosu o priniatii v zakonodatel'nom poriadke mer."

5. *Statisticheskii atlas goroda Moskvy* (1890), VI–VII; Bradley, *Muzhik and Muscovite*, 196; Mazanik, "City as a Transient Home," 65.

6. Zviaginskii, *Domovaia kanalizatsiia*, 3.

7. *Statisticheskii atlas goroda Moskvy* (1890), XX.

8. For the detailed discussion of this argument, see Mazanik, "City as a Transient Home," 61–64.

9. *Sobranie uzakonenii i rasporiazhenii pravitel'stva za 1912 god*, part 1, no. 97, June 5, 1912, law 842, pp. 1709–11; Nikitin, "Kanalizatsiia goroda Moskvy," 328. For municipal discussions about mandatory connection to the sewerage system, see also *Izvestiia Moskovskoi Gorodskoi Dumy* 4 (1906): 7–10.

10. *Izvestiia Moskovskoi Gorodskoi Dumy* 4 (1906): 1–10.

11. *Izvestiia Moskovskoi Gorodskoi Dumy* 5 (1905): 52–56.

12. *Izvestiia Moskovskoi Gorodskoi Dumy* 4 (1906): 3; Zviaginskii, *Domovaia kanalizatsiia*, 3

13. Zviaginskii, *Domovaia kanalizatsiia*, 118 (also 84–85).

14. Nikitin, "Kanalizatsiia goroda Moskvy," 329; Zviaginskii, *Domovaia kanalizatsiia*, 3; Sytin, *Kommunal'noe khoziaistvo*, 114.

15. Tarr, "From City to Farm," 608; Sheail, "Town Wastes," 203, 209; Uekoetter, "City Meets Country," 93.

16. Johnson, "Peasant and Proletariat," 82–83.

17. *Zhurnaly Komissii* (1899), 34–36.

18. *Zhurnaly Komissii* (1899), 43–44.

19. Elina, *Ot tsarskikh sadov do sovetskikh polei*, 1:298–99.

20. Stanchevici, *Stalinist Genetics*, 144.

21. *Zhurnaly Komissii* (1899), 55.

22. *Zhurnaly Komissii* (1899), 56.

23. Kotsonis, *Making Peasants Backward*, 4. See also Kopsidis, Bruisch, and Bromley, "Where Is the Backward Russian Peasant?"

24. *Zhurnaly Komissii* (1899), 45.

25. *Zhurnaly Komissii* (1899), 25.

26. Vil'iams, "Obshchie osnovaniia obezvrezhivaniia nechistot"; Vil'iams "Deiatel'nost' polei orosheniia"; Nikitin, "Kanalizatsiia goroda Moskvy," 305, 318; *Poiasnitel'naia zapiska k proektu kanalizatsii goroda Moskvy*, 142.

27. Uekoetter, "City Meets Country," 100.

28. Vil'iams, "Deiatel'nost' polei orosheniia," 171.

29. Nikitin, "Kanalizatsiia goroda Moskvy," 334–35, 344–45.

30. Nikitin, "Kanalizatsiia goroda Moskvy," 306–7.

31. Filtzer, *Hazards of Urban Life*, 30.

Chapter 5. The Conundrum of Industrial Waste

1. TsGA Moskvy, 16:133:250:14–16.

2. Petunnikov, "Materialy dlia izucheniia Moskvy"; Kastal'skii, *Moskva v sovremennom sostoianii*; Kolokolov, *Vody Sankt-Peterburga*; Kotsyn, *Opyt sistematicheskikh nabliudenii*, 157–63; Sviatlovskii, *O fabrichnykh otbrosakh*.

3. Kupriianova, "Rabochii vopros."

4. "Ustav vrachebnyi," in *Polnyi svod zakonov Rossiiskoi Imperii* (Saint Petersburg, 1912), vol. 13 §§658, 666.

5. "Ustav o promyshlennosti," in *Polnyi svod zakonov Rossiiskoi Imperii* (Saint Petersburg, 1912), vol. 11, §68. See also "Ustav stroitel'nyi," in *Polnyi svod zakonov Rossiiskoi Imperii* (Saint Petersburg, 1912), vol. 12, §165. The emergence of these norms is discussed by Owen in *Capitalism and Politics in Russia*.

6. TsGA Moskvy, 143:1:385:12.

7. "Ustav o nakazaniiakh, nalagaemykh mirovymi sud'iami," in *Polnyi svod zakonov Rossiiskoi Imperii* (Saint Petersburg, 1912), vol. 15, §§52, 53.

8. "Ustav o nakazaniiakh, nalagaemykh mirovymi sud'iami," in *Sudebnye ustavy 20 noiabra 1864 g.* (Saint Petersburg: Gosudarstvennaia kantseliariia, 1867), §111.

9. "Ustav o nakazaniyakh, nalagaemykh mirovymi sud'iami" in *Polnyi svod zakonov Rossiiskoi Imperii* (Saint Petersburg, 1912), vol. 15, §111.

10. Miller, *K otsenke sposobov ochistki*; Drozdov, *Nekotorye dannye*.

11. *Trudy Komissii*, 2:570–604.

12. Brüggemeier, "Nature Fit for Industry"; Cioc, *The Rhine*, 110–12.

13. Garcier, "Placing of Matter."

14. Melosi, *Sanitary City* (2008), 145.

15. The River Pollution Prevention Act (London: Knight & Co., 1876), 19.

16. Rosenthal, *River Pollution Dilemma*, 22–24; Calvert, "Eighth Report."

17. For the discussion of these factors in the Saratov cholera epidemic, see Henze, *Disease, Health Care and Government*, 68–75.

18. Frenkel, *Kholera i nashi goroda*, 23, 27; Henze, *Disease, Health Care and Government*, 52; Sahadeo, "Epidemic and Empire."

19. *Otchet Moskovskoi gorodskoi*.

20. TsGA Moskvy, 16:133:250:11–13.

21. TsGA Moskvy, 16:132:40:1.

22. Koz'min, *Zubatov i ego korrespondenty*, 61.

23. TsGA Moskvy, 16:133:250:32.

24. TsGA Moskvy, 16:133:250:16.

25. TsGA Moskvy, 16:133:250:1–4, 33–34, 119–20.

26. TsGA Moskvy, 16:133:250:32–33, 60–61.//
27. *Zhurnaly Komissii* (1902), 228.
28. TsGA Moskvy, 143:1:385:12.
29. *Otchet Komissii po ochistke stochnykh vod*, 13–19.
30. On the Moscow Exchange Committee, see Owen, "Doing Business in Merchant Moscow."
31. TsGA Moskvy, 143:1:385:2.
32. TsGA Moskvy, 143:1:385:3.
33. TsGA Moskvy, 143:1:385:17–18.
34. Goldberg, "Association of Industry and Trade," 371–75.
35. TsGA Moskvy, 143:1:385:58–65.
36. Brüggemeier, "Nature Fit for Industry," 38–39; Cioc, *The Rhine*, 126–29.
37. *Otchet ekspertnoi komissii*.
38. *Otchet vremennogo komiteta po izyskaniiu mer* (1913); *Otchet vremennogo komiteta po izyskaniiu mer* (1914).
39. Goldberg, "Association of Industry and Trade," 385.
40. *Sbornik dekretov i postanovlenii po narodnomu khoziaistvu*, 21–23.

Chapter 6. Feeding Moscow

1. Naumov, *O pitatel'nykh veshchestvakh*, 145.
2. Brantz, "Animal Bodies," 193–94.
3. Bliokh, *Issledovaniia po voprosam*, 45–47; Dobroslavin, *O sravnitel'noi stoimosti uluchshennoi pishchi*, 8–11; Erisman, "Pishchevoe dovol'stvo rabochikh."
4. Cronon, *Nature's Metropolis*, 207–63; Baics, *Feeding Gotham*, 57–93; Baics and Thelle, "Introduction."
5. The 1900 livestock census counted 154 million head of livestock in the Russian Empire, including 26 million horses, 43.5 million head of cattle, 70.5 million sheep, and 14 million pigs. The Russian Empire had the world's largest number of horses and the second largest number of cattle, after the United States. See *Vremennik Tsentral'nogo Statisticheskogo Komiteta MVD*, 3, and Kriukov, *Miaso i miasnye zhivotnye*, 107–8.
6. Cronon, *Nature's Metropolis*, 218–24.
7. Gorbunov, *Moskovskie gorodskie boini*, 45; *Otchet veterinarnogo upravleniia MVD za 1903 g.*, 83; *Moskva kak potrebitel'skii tsentr miasnykh produktov*, 3.
8. *Otchet o sostoianii narodnogo zdraviia* (1876), 92–93; *Issledovanie sovremennogo sostoianiia skotovodstva v Rusii*.
9. Moon, *Plough That Broke the Steppe*, 11–13; Khodarkovsky, *Russia's Steppe Frontier*, 7–33.
10. Moon, *Plough That Broke the Steppe*, 17.
11. Pronshtein, *Don i stepnoe Predkavkazie*, 93–94; O'Rourke, *Warriors and Peasants*, 65.

12. *Otchet o sostoianii narodnogo zdraviia* (1876), 93–94; Pronshtein, *Don i stepnoe Predkavkazie*, 93–100, 107–11, 122–26; Moon, *Plough That Broke the Steppe*, 18–19.

13. *Polnoe sobranie zakonov Rossiiskoi Imperii, 1825–1881*, vol. 36 (1861), part 2 (Saint Petersburg: Tipografiia II otdeleniia Sobstvennoi E.I.V. Kantseliarii, 1863), no. 37265, and vol. 42 (1867), part 2 (Saint Petersburg: Tipografiia II otdeleniia Sobstvennoi E.I.V. Kantseliarii, 1871), no. 45017; Bliokh, *Issledovaniia po voprosam*, 63–65; Bogdanov, *Miasnoi vopros v Rossii*, 6–7; Bakhtiiarov, *Briukho Peterburga*, 19.

14. *Otchet Meditsinskogo Departamenta za 1880 g.*, 333–46.

15. *Otchet veterinarnogo otdeleniia MVD za 1885 g.*, 43; *Otchet veterinarnogo upravleniia MVD za 1903 g.*, 83; Kriukov, *Miaso and miasnye zhivotnye*, 111.

16. Bakhtiiarov, *Briukho Peterburga*, 15–19.

17. Bliokh, *Issledovaniia po voprosam*, 69–71.

18. In the 1870s and early 1880s, the average dressed weight of cattle in Moscow was about 300 kilograms. See Kriukov, *Miaso i miasnye zhivotnye*, 113.

19. Swabe, *Animals, Disease and Human Society*, 10–13, 85–117. In 1873, Russia reorganized its veterinary schools, first created in the early 1800s, and established three veterinary institutes—in Dorpat (Tartu), Kharkov (Kharkiv), and Kazan', joined in 1889 by another institute in Warsaw. See Tatarskii, "Veterinarnye instituty i shkoly."

20. Ia. P[olferov]. "Chuma rogatogo skota," 52–54.

21. John Gamgee of Albert Veterinary College, one of the most vocal British veterinarians of the time, connected the outbreak to a shipment of cattle from the port of Revel (Tallinn) in the Russian Empire in August 1865. See Erickson, "Cattle Plague in England"; Swabe, *Animals, Disease and Human Society*; Worboys, *Spreading Germs*, 49–60.

22. Wilkinson, *Animals and Disease*, 104; Veselovskii, *Istoriia zemstva*, 2:366.

23. For the history of cattle plague study and control in the Kazakh steppe, see Duisebayeva, "Animal Face of Imperial Power," 59–63.

24. *Polnyi Svod Zakonov Rossiiskoi Imperii. Sobranie vtoroe, 1825–1881*, vol. 43, part 2 (Saint Petersburg, 1873), §46503.

25. *Polnyi Svod Zakonov Rossiiskoi Imperii. Sobranie vtoroe, 1825–1881*, vol. 45, part 2 (Saint Petersburg, 1874), §48576.

26. *Polnyi Svod Zakonov Rossiiskoi Imperii. Sobranie vtoroe, 1825–1881*, vol. 51, part 2 (Saint Petersburg, 1878), §56634.

27. *Polnyi Svod Zakonov Rossiiskoi Imperii. Sobranie vtoroe, 1825–1881*, vol. 54, part 1 (Saint Petersburg, 1881), §59739; Veselovskii, *Istoriia zemstva*, 2:361–72.

28. For the discussion of these measures, see Mazanik, "Public Health across Species."

29. Ia. P[olferov], "Chuma rogatogo skota," 53–54.

30. Bogdanov, *Miasnoi vopros v Rossii*, 7–8.

31. Cronon, *Nature's Metropolis*, 223–24.

32. Kriukov, *Miaso i miasnye zhivotnye*, 114–17; Bogdanov, *Miasnoi vopros v Rossii*, 7–21.

33. Zhbankov, "Kratkie svedeniia," 45–46; Veselovskii, *Istoriia zemstva za sorok let*, 2:407; Mazanik, "Public Health across Species."

34. Veselovskii, *Istoriia zemstva*, 2:412–14. Elizabeth Hachten analyzes the controversies related to the early anthrax vaccine application in Russia in "Science in the Service of Society," 211–14, 221–25, and "In Service to Science and Society," 198–200.

35. *Sbornik postanovlenii*, 265; Veselovskii, *Istoriia zemstva*, 2:419; Mazanik, "Public Health across Species."

36. Wilkinson, "Glanders," 383.

37. Iavorskii, *Gorodskoi veterinarnyi nadzor*, 5–11.

38. TsGA Moskvy, 179:58:50:62–75; Iavorskii, *Gorodskoi veterinarnyi nadzor*, 2–3.

39. Iavorskii, *Gorodskoi veterinarnyi nadzor*, 5.

Chapter 7. The Killing Factory

1. Joyce, *Rule of Freedom*, 76–83; Otter, "Civilizing Slaughter," 30.

2. Vialles, *Animal to Edible*, 22.

3. For the discussion of the term *abattoir*, see Vialles, *Animal to Edible*, 15–26; Joyce, *Rule of Freedom*, 77.

4. Joyce, *Rule of Freedom*, 76–77; Young Lee, *Meat, Modernity and the Rise of the Slaughterhouse*.

5. "Doklad N 41," appendix 5, 22.

6. Watts, "Liberty, Equality and the Public Good," 117.

7. Poderni, *Tekhnicheskoe opisanie*, iv.

8. "Zapiska o rabotakh," 1–2.

9. Gorbunov, *Moskovskie gorodskie boini*, 20–21.

10. The proposals of the reorganization and centralization of slaughter can be seen in *Izvestiia Moskovskoi Gorodskoi Dumy* 3 (1885): 13. See also Verner, "Moskovskii skotnyi i miasnoi rynok," 37.

11. Brantz, "Animal Bodies," 199. See also Virkhov, *Izlozhenie ucheniia o trikhinakh*.

12. Rudnev, *O trikhinakh v Rossii*, 1–2, 24.

13. Chudnovskii, *Vorpos o trikhnakh*; Andreievskii, *Glisty i trikhiny*; Belin, "Demonstrirovanie trikhin"; Tikhomirov, *O legchaishem sposobe otkrytiia trikhin*; Krylov, *K istorii trichinoza v Rossii*; Zeyfman, *Trikhiny i trikhinnaia bolezn'*.

14. Brantz, "How Parasites Make History."

15. Brantz, "How Parasites Make History," 72.

16. *Donesenie Meditsinskomu sovetu*; Petropavlovskii, *K voprosu o rasprostranenii trikhin*, 12–14.

17. "Doklad N 41," 1–2.

18. On Russian "bimetropolitanism" and the rivalry between Moscow and Saint Petersburg, see Shevyrev, "Axis Petersburg-Moscow." On the Moscow bourgeoisie, see Rieber, *Merchants and Entrepreneurs*; Petrov, *Moskovskaia burzhuaziia*.

19. "Doklad N 41," 37–38, 80.

20. "Doklad N 41," 9–10, 29.

21. "Doklad N 41," 12–14.

22. "Doklad N 41," 32; Verner, *Sovremennoe khoziaistvo*, 243.

23. Minutes of the city council discussion were published in *Izvestiia Moskovskoi Goordoskoi Dumy* 7 (1885): 775, also 774–91.

24. *Izvestiia Moskovskoi Gorodskoi Dumy* 10 (1887): 860.

25. For the US case, see Cronon, *Nature's Metropolis*, 207–59.

26. "Doklad N 41," 19.

27. *Izvestiia Moskovskoi Gorodskoi Dumy* 7 (1885): 787.

28. *Izvestiia Moskovskoi Gorodskoi Dumy* 10 (1885): 1113.

29. "Doklad N 41," appendix 5, 15.

30. Sunier and Ewen, *Another Global City*.

31. *Izvestiia Moskovskoi Gorodskoi Dumy* 7 (1885): 787–88; *Izvestiia Moskovskoi Gorodskoi Dumy* 10 (1885): 1109.

32. *Izvestiia Moskovskoi Gorodskoi Dumy* 7 (1885): 781.

33. Maclachlan, "Bloody Offal Nuisance."

34. Poderni, *Tekhnicheskoe opisanie*, v, 4; Gorbunov, *Moskovskie gorodskie boini*, 28.

35. Gorbunov, *Moskovskie gorodskie boini*, 36 (also 33–36).

36. Poderni, *Tekhnicheskoe opisanie*, 19–83; Smolenskii, *Boini i skotoprigonnye dvory*, 11–12.

37. Poderni, *Tekhnicheskoe opisanie*, 25, 55, 126–37; *Shone Hydro-Pneumatic System of Sewerage*, 39–47.

38. Poderni, *Tekhnicheskoe opisanie*, 34–41.

39. *Veterinarnyi nadzor*, 6.

40. Nagorskii, "Veterinarny nadzor," 4–5; V. F[idler], *Moskva*, 88.

41. Gorbunov, *Moskovskie gorodskie boini*, 45; *Moskva kak potrebitel'skii tsentr myasnykh produktov*. 3.

42. "Ustav vrachebny," in *Svod zakonov Rossiiskoi Imperii* (Saint Petersburg: S.n., 1892), vol. 13, appendix to article 1265, 271.

43. "Otnoshenie veterinarnogo vracha K.Z. Kleptsova," 15.

44. Gorbunov, *Moskovskie gorodskie boini*, 60–61.

45. TsGA Moskvy, 179:54:947:20.

46. Gorbunov, *Moskovskie gorodskie boini*, 62–63. See also Iavorskii, *Gorodskoi veterinarnyi nadzor*, 11.

47. Koch, "Address on the Transference"; Hime, "Shall We Eat Tuberculous Meat?"; Gorbunov, *Moskovskie gorodskie boini*, 62.

48. Gorbunov, *Moskovskie gorodskie boini*, 63.

49. *Veterinarnyi nadzor*, 7–8.

50. "Doklad N 41," appendix 5, 22.

51. Zelenin, "Moskovskie gorodskie boini," 498–99; alsoTsGA Moskvy, 179:54:992:92–97.

52. Zelenin, "Moskovskie gorodskie boini," 479.

53. Gorbunov, *Moskovskie gorodskie boini*, 58–60.

54. TsGA Moskvy, 179:54:992:109.

55. TsGA Moskvy, 179:54:1057, 1105, 980; Zelenin, "Moskovskie gorodskie boini," 500.

56. *Illiustrirovannyi putevoditel' po Moskve*, 96; *Moskva. Putevoditel'*, 260; *Sputnik moskvicha*, 79.

Chapter 8. Civilized Slaughter

1. Pearson, "Speaking Bodies," 91–92; also Shaw, "Way with Animals," 2–3.

2. For discussions on the humane treatment of animals in Victorian Britain, see James Turner, *Reckoning with the Beast*.

3. *Ustav Rossiiskogo Obshchestva* (1865), 1–2. The clause about common people was removed in the later version of the charter. See *Ustav Rossiiskogo Obshchestva* (1888).

4. *Polnoe Sobranie Zakonov Rossiiskoi Imperii*, ser. 2, vol. 46 (1871), part 2 (Saint Petersburg: S.n., 1874), no. 50208, §43; also Nelson, "Body of the Beast."

5. *Otchet o deiatel'nosti Rossiiskogo Obshchestva Pokrovitel'stva Zhivotnym za 1891 god*, 5, 44–45; *Otchet o deiatel'nosti Rossiiskogo Obshchestva Pokrovitel'stva Zhivotnym za 1901 god*, 3–4.

6. *Otchet o deiatel'nosti Rossiiskogo Obshchestva Pokrovitel'stva Zhivotnym za 1891 god*, 3.

7. *Ustav Rossiiskogo Obshchestva* (1888), 1–2.

8. TsGA Moskvy, 179:54:992:128–30.

9. For the study of Tolstoy's vegetarianism, see LeBanc, "Tolstoy's Way of No Flesh."

10. TsGA Moskvy, 179:54:992:130. The general activities of the RSPA also included the creation of horse abattoirs and the intentional purchase of old horses for slaughter. See *Ustav Rossiiskogo Obshchestva* (1888), 2–3.

11. TsGA Moskvy, 179:54:992:130.

12. TsGA Moskvy 179:54:992:131.

13. TsGA Moskvy 179:54:992:134–35.

14. For a brief discussion of these responses in Russian, see Mazanik, "Vospriiatie nasiliia."

15. TsGA Moskvy 179:54:992:137.

16. TsGA Moskvy 179:54:992:141.

17. TsGA Moskvy 179:54:992:144.

18. This interpretation relies on the arguments of Temple Grandin, an American animal scientist who claims that her own autism allows her to better understand animals and thus able to redesign slaughterhouses and other livestock handling facilities to improve the experience of the animals, which she has done with huge success. See Grandin, "Making Slaughterhouses More Humane"; Fudge, "Milking Other Men's Beasts," 18–19.

19. TsGA Moskvy 179:54:992:145.

20. TsGA Moskvy 179:54:992:142. The catastrophe of Khodynka refers to the human stampede that occurred on Khodynka Field in Moscow during the festivities connected to the coronation of Nicholas II in 1896, which caused numerous deaths.

21. TsGA Moskvy, 179:54:992:138.

22. TsGA Moskvy 179:54:992:150.

23. TsGA Moskvy 179:54:992:151–52.

24. TsGA Moskvy 179:54:992:152–53.

25. Zelenin, "Moskovskie gorodskie boini," 498–99. These numbers do not include employees at the industrial plants located at the abattoir but not under municipal control.

26. TsGA Moskvy, 179:54:982:1.

27. Zelenin, "Moskovskie gorodskie boini," 478.

28. Gorbunov, *Moskovskie gorodskie boini*, 71; Zelenin, "Moskovskie gorodskie boini," 478–79.

29. TsGA Moskvy 179:54:995:6, 179:54:1054:3–5; *Izvestiia Moskovskoi Gorodskoi Dumy* (June–July 1905): 25–26.

30. TsGA Moskvy, 179:54:1068:3–7; Gorbunov, *Moskovskie gorodskie boini*, 79.

31. TsGA Moskvy, 179:54:1112:34–37.

32. TsGA Moskvy, 179:54:1112:36–37, 87–89.

33. Bater, *St Petersburg*, 287–95; Dement'iev, *Fabrika*, 36–43.

34. TsGA Moskvy, 179:54:1054:7–8; Poderni, *Tekhnicheskoe opisanie*, 83–99.

35. TsGA Moskvy, 179:54:1112:27, 90A, 179:54:1069:12.

36. TsGA Moskvy, 179:54:1106:1, 179:54:1068:4, 179:54:1054:7; Zelenin, "Moskovskie gorodskie boini," 500.

37. TsGA Moskvy, 179:54:631:63; Poderni, *Tekhnicheskoe opisanie*, 95.

38. TsGA Moskvy, 179:54:631:2; *Izvestiia Moskovskoi Gorodskoi Dumy* (June–July 1905): 13–23.

39. TsGA Moskvy, 179:54:631:63–74; *Izvestiia Moskovskoi Gorodskoi Dumy* (June–July 1905): 25–7.

40. *Izvestiia Moskovskoi Gorodskoi Dumy* (October 1905): 47–48. See also Engelstein, *Moscow, 1905*, 21, 116–20.

41. TsGA Moskvy, 179:54:631:133, 179:54:632:16–17; *Izvestiia Moskovskoi Gorodskoi Dumy* (November 1905): 38–48; Gorbunov, *Moskovskie gorodskie boini*, 47.

42. *Izvestiia Moskovskoi Gorodskoi Dumy* (November 1905): 39.

Chapter 9. A Deadly City for Children

1. Bubnov's report is quoted in Nikitenko, *Detskaia smertnost'*, 9–10.

2. Byford, *Science of the Child*, 1–6; Byford, "Roditel', uchitel' i vrach"; Gorshkov, *Russia's Factory Children*, 123–27.

3. Kelly, *Children's World*, 26–28; Gorshkov, *Russia's Factory Children*, 128–65.

4. See Ransel, *Mothers of Misery*; Kelly, *Children's World*; Chernyaeva, "Childcare Manuals"; White, *Modern History of Russian Childhood*; also Kolganova, "Zarozhdenie sistemy."

5. Kurkin, "Sanitarnaia statistika," 90 (mortality statistics for 1898–1904); Sokolov and Grebenshchikov, *Smertnost' v Rossii*, 21–24, 53.

6. Sokolov and Grebenshchikov, *Smertnost' v Rossii*, 4–8, 36–45; Kelly, *Children's World*, 293–98; Chernyaeva, "Childcare Manuals," 28–32.

7. Nikitenko, *Detskaia smertnost'*, 10–11.

8. Nikitenko, *Detskaia smertnost'*, 94–99, 230–31.

9. Johnson, "Peasant and Proletarian," 82–83; Engel, *Between the Fields and the City*, 82–85.

10. Johnson, "Peasant and Proletarian," 85–87; Engel, *Between the Fields and the City*, 127; Engel, "The Woman's Side"; Bradley, *Muzhik and Muscovite*, 137.

11. *Glavenishie predvaritel'nye dannye perepesi g. Moskvy 1902 g. Naselenie Moskvy po zaniatiiam* (Moscow: Gorodskaia Tipografiia, 1907), 32, 55.

12. *Tablitsy o dvizhenii naseleniia v g. Moskve v 1890 g.*, part I, 3–18.

13. Engel, *Between the Fields and the City*, 126–28.

14. *Smertnost' naseleniia*, 43–44.

15. Ransel, *Mothers of Misery*, 3.

16. *Smertnost' naseleniia*, 43–44.

17. Sutugin, *Kratkii ocherk*, 9.

18. Iablokov, *Prizrenie detei*, 45–51, 65.

19. *Smertnost' naseleniia*, 41–48.

20. Ransel, *Mothers of Misery*,167; Iablokov, *Prizrenie detei*, 59–65.

21. *Statisticheskii ezhegodnik g. Moskvy i Moskovskoi gubernii*, part III, 93.

22. *Tablitsy o dvizhenii naseleniia v g. Moskve v 1890 g.*, part II, 1–12; *Statisticheskii ezhegodnik g. Moskvy i Moskovskoi gubernii*, part III, 104.

23. See, for example, the title of chapter 9 in Ruble, *Second Metropolis*, 265–91.

24. Mazanik, "City as a Transient Home," 63–64. The statistics are from "Doklad Moskovskoi gorodskoi upravy, 1914," quoted in Alaverdian, *Zhilishchnyi vopros v Moskve*, 33–34. The same statistics are also used in studies by Bradley, *Muzhik and Muscovite*, 196, and Ruble, *Second Metropolis*, 266. For comparisons with Saint Petersburg, see Bater, *St Petersburg*, 329.

25. Verner, *Zhilishcha bedneishego naseleniia*; Kishkin, "Zhilishchnyi vopros v Moskve"; *Statisticheskii ezhegodnik g. Moskvy*, 82.

26. *Sanitarnye usloviia koechno-komorochnykh kvartir*, 9–10.

27. Gornostaev, *Deti rabochikh*, 5 (also 4–8).

28. *WHO Housing and Health Guidelines* (Geneva: World Health Organization, 2018), https://www.ncbi.nlm.nih.gov/books/NBK535289/.

29. *Smertnost' naseleniia*, tables, 15; *Statisticheskii ezhegodnik g. Moskvy i Moskovskoi gubernii*, Part III, 104.

30. *Statisticheskii ezhegodnik g. Moskvy i Moskovskoi gubernii*, Part III, 93.

31. Uspenskii, *Moskva*, 53.

32. *Ocherk vozniknoveniia*, 10–23.

33. Chernyaeva, "Childcare Manuals," 52.

34. Chernyaeva, "Childcare Manuals," 37–42.

35. Uspenskii, *Moskva*, 52.

36. Medovikov, *V chem dolzhna sostoiat' bor'ba*, 13–17; Verner, *Sovremennoe khoziaistvo*, 138–43.

37. La Berge, "Medicalization and Moralization."

38. Lindenmayer, *Poverty Is Not a Vice*, 148–54.

39. Kamenetskaia, *Blagotvoritel'naia deiatel'nost'*, 23; Gornostaev, *Deti rabochikh i*, 22–28.

40. Lindenmayer, *Poverty Is Not a Vice*, 150.

41. La Berge, "Medicalization and Moralization."

42. *Izvestiia Moskovskoi Gorodskoi Dumy* 17 (1905): 67; Verner, *Sovremennoe khoziaistvo*, 222.

43. *Statisticheskii ezhegodnik g. Moskvy i Moskovskoi gubernii*, part III, 89.

44. Chernyaeva shows that the Moscow Charitable Society for the Protection of Motherhood had facilities for only about 430 children in 1915; the All-Russian Guardianship for the Protection of Motherhood and Infancy had 29 urban crèches for 900 children in 1917. See Chernyaeva, "Childcare Manuals," 52–54.

45. *Mezhduvedomstvennaia komissiia po peresmotru*.

46. Kishkin, "Zhilishchnyi vopros v Moskve," 292.

47. Uspenskii, *Moskva*, 56–58.

48. Verner, *Sovremennoe khoziaistvo*, 188, 192–95.

49. Bogopol'skii, "K voprosu o priniatii"; Pokrovskaia, *Sanitarnyi nadzor nad zhilishchami*, 128, 132.

50. Thurston, *Liberal City*, 140–41.

51. Verner, *Sovremennoe khoziaistvo*, 178–79.

52. Verner, *Sovremennoe khoziaistvo*, 189–92.

53. *Otchet o sostoianii narodnogo zdraviia* (1903), 15–16; *Otchet o sostoianii narodnogo zdraviia* (1904), 15–16; *Otchet o sostoianii narodnogo zdraviia* (1905), 10–12; *Otchet o sostoianii narodnogo zdraviia* (1906), 10–12; *Otchet of sostoianii narodnogo zdraviia* (1912), 9.

54. *Statisticheskii ezhegodnik g. Moskvy i Moskovskoi gubernii*, part III, 89; Starks, *Body Soviet*, 138–56; Goldman, "Women, Abortion, and the State," 249–60, 262–63; Chernyaeva, "Childcare Manuals," 86–91.

55. Sytin, *Kommunal'noe khoziaistvo*, 78.

Chapter 10. Healthy Schools in a Deadly City

1. Eklof, *Russian Peasant Schools*, 287–99. Eklof suggests that the actual number of boys and girls who received some schooling was much higher than was reported in the official statistics.

2. Eklof, *Russian Peasant Schools*; Kelly, *Children's World*; Byford, *Science of the Child*.

3. Erisman, *Vliianie shkol*, 1.

4. Cohn, *Untersuchungen der Augen*; Becker, *Luft und Bewegung*; Guillaume, *Hygiène scolaire*.

5. Virchow, *Ueber gewisse die Gesundheit*.

6. Virkhov, "O vrednykh vliianiiakh shkoly na zdorovie."

7. Meckel, *Classrooms and Clinics*, 13–14.

8. Erisman, *Vliianie shkol*, 80; Erisman, *Professional'naia gigiena*, 29; Virenius, *Organizatsiia sanitarnoi chasti*, 42; Vasilievskii, *Gigiena i sanitariia*, 50–51; V. M. O-ii [Ostrovskii], "Shkol'naia gigiena."

9. Nagorskii, *O vliianii shkol*, 4, 10, 14, 27–37. On the link between chest girth and predisposition to consumption, see Verner, *Sovremennoe khoziaistvo*, 61.

10. Erisman, *Vliianie shkol*, 98. See also Virenius, *Shkol'nye stoly i skam'i*.

11. Erisman, *Vliianie shkol*; Erisman, *Soobrazheniia po ustroistvu obraztsovoi klassnoi komnaty*, 4–9; Erisman, *Soobrazheniia po voprosu o nailuchshem ustroistve klassnoi mebeli*, 3–4. The latter was published in the official circular letter for the Moscow educational district, which suggests that Erismann's recommendations were recognized at the high level.

12. Pavlov, *Narodnaia shkola*, 1.

13. Meckel, *Classrooms and Clinics*, 12, 31–36.

14. Korf, *Russkaia nachal'naya shkola*, 72–78.

15. Zavolzhskaia, *Shkol'naia gigiena*, 82. See also Levenson, *Gigiena shkol*, 28.

16. Levenson, *Gigiena shkol*, 19–23; Erisman, *Professional'naia gigiena*, 30.

17. Erisman, *Professional'naia gigiena*, 30–31.

18. Pavlova, *Sbornik podvizhnykh igr*, introduction by F. Erismann, 1–2.

19. Eklof, "Worlds in Conflict."
20. Erisman, *Professional'naia gigiena*, 32.
21. Virenius, *Organizatsiia sanitarnoi chasti*, 40 (also 37–40).
22. Meckel, *Classrooms and Clinics*, 37, 52–61; Hofmann, *Gesundheitswissen in der Schule*, 116–17.
23. Meckel, *Classrooms and Clinics*, 37.
24. *Nachal'nye uchilishcha*, 1.
25. Krasnopevkov, "Obzor uchilishch."
26. Krasnopevkov, "Obzor uchilishch."
27. *Moskovskie gorodskie nachal'nyie uchilishcha*, 2; Verner, *Sovremennoe khoziaistvo*, 33–36.
28. Verner, *Sovremennoe khoziaistvo*, 36–38.
29. *Moskovskie gorodskie nachal'nyie uchilishcha*, 7–14, 45–46.
30. TsGA Moskvy, 179:56:95:13.
31. Mikhailov, *Materialy k opredeleniiu*; Mikhailov, *Obshchaia kharakteristika deiatel'nosti*.
32. TsGA Moskvy, 179:56:95:14–15.
33. TsGA Moskvy, 179:56:105:9–11.
34. TsGA Moskvy, 179:56:105:10.
35. Byford, "Professional Cross-Dressing," 598.
36. Byford, "Professional Cross-Dressing," 595.
37. Eklof, *Russian Peasant Schools*; Eklof, "Worlds in Conflict"; Eklof and Petersen, "Laska i Poriadok"; Foucault, *Discipline and Punish*.
38. Eklof, "Worlds in Conflict," 795. Catriona Kelly suggests that this argument has its limitations, and that rigorous inspections by the Ministry of Education and the demands of academic curricula imposed considerable constraints on the "noncoercive" and "child-centered" classroom, especially in secondary schools. See Kelly, *Children's World*, 33.
39. TsGA Moskvy, 179:56:105:14.
40. Mikhailov, *Materialy k opredeleniiu*, 23.
41. TsGA Moskvy, 179:56:105:47.
42. TsGA Moskvy, 179:56:105:156.
43. "O zavtrakakh v nachal'nykh uchilishchakh goroda Moskvy," 38.
44. TsGA Moskvy, 179:56:105:161.
45. *Izvestiia Moskovskoi Gorodskoi Dumy, Vrachebno-sanitarnyi otdel* (March 1903): 16.
46. Verner, *Sovremennoe khoziaistvo*, 63–64.
47. "O sanitarnom sostoianii pomeshchenii"; TsGA Moskvy, 179:56:105:52, 179:56:119:52–56.
48. "O zavtrakakh v nachal'nykh uchilishchakh," 44.
49. Verner, *Sovremennoe khoziaistvo*, 39–40.

50. *Izvestiia Moskovskoi Gorodskoi Dumy, Vrachebno-sanitarnyi otdel* (October 1903): 5; Verner, *Sovremennoe khoziaistvo*, 133.

51. TsGA Moskvy, 179:56:119:132–34.

52. Meckel, *Classrooms and Clinics*, 52–61.

53. TsGA Moskvy, 179:56:105:56.

54. TsGA Moskvy, 179:56:263:1.

55. Verner, *Sovremennoe khoziaistvo*, 71–73.

56. *Moskovskaia gorodskaia uprava. Obzor po gorodu Moskve za 1910 g.*

57. *Izvestiia Moskovskoi Gorodskoi Dumy* 9 (1905): 45–46.

58. TsGA Moskvy, 179:56:263:3.

59. *Letnie kolonii. Otchet 1890 g.*, 12; Verner, *Sovremennoe khoziaistvo*, 67.

60. *Letnie kolonii. Otchet 1896 g.*, 12.

61. *Entsiklopedicheskii slovar' Brokgauza-Efrona*, vol. 39a (1903), 659.

62. *Letnie kolonii. Otchet 1891 g.*, 11. See also *Letnie kolonii. Otchet 1890 g.*, 4–12; *Letnie kolonii. Otchet 1891 g.*, 7–13; *Letnie kolonii. Otchet 1893 g.*, 20.

63. *Letnie kolonii. Otchet 1893 g.*, 17. See also *Letnie kolonii. Otchet 1891*, 11; *Letnie kolonii. Otchet 1892 g.*, 5.

64. *Letnie kolonii. Otchet 1892 g.*, 5; *Letnie kolonii. Otchet 1893 g.*, 13–15.

65. *Letnie kolonii. Otchet 1891 g.*, 15 (also 15–19).

66. *Letnie kolonii. Otchet 1890 g.*, 4; *Letnie kolonii. Otchet 1891 g.*, 16–20; *Letnie kolonii. Otchet 1893 g.*, 4; Verner, *Sovremennoe khoziaistvo*, 67.

67. Byford, "Professional Cross-Dressing," (esp. 614).

Conclusion: Across the Divide

1. Shchepkin, *Obshchestvennoe samoupravlenie*, 44.

2. Shchepkin, *Obshchestvennoe samoupravlenie*, 46.

3. Hutchinson, *Politics and Public Health*, 69–72; Vigdorchik, "Voprosy narodnogo zdraviia," 13; Strashun, *Russkaia obshchestvennaia meditsina*, 91–92.

4. *Moskovskaia gorodskaia duma, 1913–1916*, 56, 87; *Ocherk deiatel'nosti Vserossiiskogo soiuza gorodov*, 174–75; Pisar'kova, *Gorodskie reformy*, 386–87.

5. Hutchinson, *Politics and Public Health*, 108–17.

6. *Ocherk deiatel'nosti Vserossiiskogo soiuza gorodov*, 56–59, 90–100.

7. *Moskovskaia gorodskaia duma, 1913–1916*, 129–31.

8. *Deiatel'nost' Moskovskogo gorodskogo upravleniia*, 20.

9. *Moskovskaia gorodskaia duma, 1913–1916*, 144.

10. *Deiatel'nost' Moskovskogo gorodskogo upravleniia*, 30, 35; *Moskovskaia gorodskaia duma, 1913–1916*, 153.

11. *Moskovskaia gorodskaia duma, 1913–1916*, 56–58.

12. For example, during the 1915 cholera outbreak in Moscow, mortality remained unusually low thanks to improved sanitation and the vaccination programs. See Davis, *Russia in the Time of Cholera*, 140–41.

13. *Statisticheskii ezhegodnik g. Moskvy i Moskovskoi gubernii*, 88–89, 94–95.

14. See, for example, the diaries of Vladimir Korolenko, Stepan Veselovskii, Tatiana Sukhotina-Tolstaia, and Nikolai Mendelson for this period, digitized by the project "Prozhito," Center for the Study of Ego-Documents, European University at Saint Petersburg, https://prozhito.org/.

15. Okunev, *Dnevnik moskvicha*, 2:13.

16. On the pharmaceutical crisis of the early Soviet years, see Zatravkin and Vishlenkova, *"Kluby" i "getto" sovetskogo zdravookhraneniia*, 104–11.

17. *Statisticheskii ezhegodnik g. Moskvy i Moskovskoi gubernii*, 88, 94.

18. Zhbankov, *Sbornik po gorodskomu vrachebno-sanitarnomu delu*, 1.

19. Filtzer, *Hazards of Urban Life*, 67.

BIBLIOGRAPHY

Archival Sources

Tsentral'nyi gosudarstvennyi arkhiv goroda Moskvy [TsGA Moskvy].
   F. 16 Moskovskii General-Gubernator.
   F. 143 Moskovskii Birzhevoi Komitet.
   F. 179 Moskovskoe Gorodskoe Upravlenie.

Cited Publications

Agafonova, Anna. "Urban Pollution and Water Supply in Novgorod, 1870–1914." *Historia Urbana* 28 (2020): 225–47.

Agafonova, Anna. "Water Supply to the Small Cities in the Northern Region of the Russian Empire, 1890–1910s (Vologda, Staraya Russa and Cherepovets)." *City and History* 9 (2020): 45–68.

Alaverdian, S. K. *Zhilishchnyi vopros v Moskve*. Yerevan: Izdatel'stvo AN Arm.SSR, 1961 [1915].

Al'bom moskovskikh gorodskikh popechitelstv o bednykh. Moscow: Gorodskaia tipografiia, 1902.

Al'bom zdanii, prinadlezhashchikh Moskovskomu gorodskomu obshchestvennomu upravleniiu. Moscow: Gorodskaia tipografiia, 191–.

Alexander, John. *Bubonic Plague in Early Modern Russia: Public Health and Urban Disaster*. Oxford: Oxford University Press, 2003.

Anderson, Barbara. *Internal Migration during Modernization in Late Nineteenth-Century Russia*. Princeton, NJ: Princeton University Press, 1980.

Andreievskii, V. *Glisty i trikhiny: Ikh proiskhozhdenie, stroenie, otlichitel'noe raspoznavanie i mikroskopicheskoe issledovanie*. Saint Petersburg: Tipografiia Golovachova, 1867.

Arnold, David. *Science, Technology and Medicine in Colonial India*. Cambridge: Cambridge University Press, 2000.

Astrakov, V. I. "O kolichestve vody, protekaiushchei v reke Moskve." *Izvestiia Moskovskoi Gorodskoi Dumy* 8 (1878): 39–41.

Bibliography

Baics, Gergely. *Feeding Gotham: The Political Economy and Geography of Food in New York, 1790–1860*. Princeton, NJ: Princeton University Press, 2016.

Baics, Gergely, and Mikkel Thelle. "Introduction: Meat and the Nineteenth-Century City." *Urban History* 45, no. 2 (2018): 184–92.

Bakhtiiarov, A. A. *Briukho Peterburga: Ocherki stolichnoi zhizni*. Saint Petersburg: Fert, 1994 [1887].

Barnes, David. *The Great Stink of Paris and the Nineteenth-Century Struggle against Filth and Germs*. Baltimore, MD: Johns Hopkins University Press, 2006.

Bater, James. *St Petersburg: Industrialization and Change*. Montreal: McGill-Queen's University Press, 1976.

Becker, Elisa. *Medicine, Law, and the State in Imperial Russia*. Budapest: Central European University Press, 2011.

Becker, Theodor. *Luft und Bewegung zur Gesundheitspflege in den Schulen*. Frankfurt am Main: Guchsland, 1867.

Beer, Daniel. *Renovating Russia: The Human Sciences and the Fate of Liberal Modernity, 1880–1930*. Ithaca, NY: Cornell University Press, 2008.

Behrends, Jan C., and Martin Kohlrausch, eds. *Races to Modernity: Metropolitan Aspirations in Eastern Europe, 1890–1940*. Budapest: Central European University Press, 2014.

Belin, M. A. "Demonstrirovanie trikhin, naidennykh v vetchine Riullinga." *Moskovskaia meditsinskaya gazeta* no. 50 (1874): 1694–95.

Benidickson, Jamie. *The Culture of Flushing: A Social and Legal History of Sewage*. Vancouver: University of British Columbia Press, 2007.

Bernstein, Laurie. *Sonia's Daughters: Prostitutes and Their Regulation in Imperial Russia*. Berkeley: University of California Press, 1995.

Bibikova, L. V. "Politicheskaia politsiia, konservatory i sotsialisty: igra liberalizmami v publichnom i nepublichnom prostranstve Rossiiskoi imperii v kontse XIX–nachale XX veka." In *Poniatiia o Rossii: K istoricheskoi semantike imperskogo perioda*, vol. 1, edited by A. Miller, D. Sdvizhkov, and I. Shirle, 514–58. Moscow: Novoe Literaturnoe Obozrenie, 2011.

Biehler, Dawn Day. *Pests in the City: Flies, Bedbugs, Cockroaches, and Rats*. Seattle: University of Washington Press, 2013.

Bliokh, I. S. *Issledovaniia po voprosam, otnosiashchimsia k proizvodstvu torgovli i peredvizheniiu skota i skotskikh produktov v Rossii i za granitsei*. Saint Petersburg: Tipografiia M. F. Volfa, 1876.

Bogdanov, E. A. *Miasnoi vopros v Rossii i sovremennoe polozhenie skoto- i miasopromyshlennosi v Rossii*. Moscow: Tipografiia Somovoi, 1912.

Bogopol'skii, B. L. "K voprosu o priniatii v zakonodatel'nom poriadke mer dlia uluchsheniia v gorodakh sanitarnogo sostoianiia vsekh zhilykh pomeshchenii, sdavaemykh vnaiem." In *Vos'moi Pirogovskii s'iezd*, vol. 3, 47–48. Moscow: S.n., 1902.

Bibliography

Bonner, Thomas Neville. "Rendezvous in Zurich: Seven Who Made a Revolution in Women's Medical Education, 1864–1874." *Journal of the History of Medicine and Allied Sciences* 44, no. 1 (1989): 7–27.

Borrero, Mauricio. *Hungry Moscow: Scarcity and Urban Society in the Russian Civil War, 1917–1921*. New York: Peter Lang, 2003.

Boubnoff, S. *Institut d'hygiène de l'Université impériale de Moscou*. Moscow: Tipo-litogragiia T-va I.I. Kushnerev i Ko, 1897.

Bradley, Joseph. *Muzhik and Muscovite: Urbanization in Late Imperial Russia*. Berkeley: University of California Press, 1985.

Brantz, Dorothee. "Animal Bodies, Human Health and the Reform of Slaughterhouses in Nineteenth-Century Berlin." *Food and History* 3, no. 2 (2005): 193–215.

Brantz, Dorothee. "How Parasites Make History: Pork and People in the Nineteenth Century." *Bulletin of the German Historical Institute Washington* 36 (2005): 69–79.

Breyfogle, Nicholas, ed. *Eurasian Environments: Nature and Ecology in Imperial Russian and Soviet History*. Pittsburgh, PA: University of Pittsburgh Press, 2018.

Brower, Daniel R. *Russian City between Tradition and Modernity*. Berkeley: University of California Press, 1990.

Brüggemeier, Franz-Josef. "A Nature Fit for Industry: The Environmental History of the Ruhr Basin, 1840–1990." *Environmental History Review* 18, no. 1 (Spring 1994): 35–54.

Burds, Jeffrey. *Peasant Dreams and Market Policies: Labor Migration and the Russian Village, 1861–1905*. Pittsburgh, PA: University of Pittsburgh Press, 1998.

Buryshkin, P. A. *Moskva kupecheskaia*. New York: Izdatel'stvo imeni Chekhova, 1954.

Butler, David, and John Davies. *Urban Drainage*. New York: Spon Press, 2011.

Byford, Andy. "Professional Cross-Dressing: Doctors in Education in Late Imperial Russia (1881–1917)." *Russian Review* 65 (October 2008): 586–616.

Byford, Andy. "Roditel', uchitel' i vrach: k istorii ikh vzaimootnoshenii v dele vospitaniia i obrazovniia v dorevolyutsionnoi Rossii." *Novyie rossiiskie gumanitarnye issledovniia* (July 2, 2013) http://www.nrgumis.ru/articles/276/.

Byford, Andy. *Science of the Child in Late Imperial and Early Soviet Russia*. Oxford: Oxford University Press, 2020.

Calvert, H. T. "The Eighth Report of the Royal Commission on Sewage Disposal." *Journal of the Society of Chemical Industry* 32, no. 6 (1913): 265–74.

Chernyaeva, Natalia. "Childcare Manuals and the Construction of Motherhood in Russia, 1890–1990." PhD diss., University of Iowa, 2009.

Chertov, A. *Gorodskaia meditsina v evropeiskoi Rossii*. Moscow: Pechatnia Iakovleva, 1903.

Chicherin, B. N. *Vospominaniia. Moskovskii universitet. Zemstvo i Moskovskaia duma*. Moscow: Izdatel'stvo im. Sabashnikovykh, 2010 [1929].

Chudnovskii, Iu. T. *Vorpos o trikhinakh i trikhinnoi bolezni v primenenii k Rossii*. Saint Petersburg: Obshchestvennaia pol'za, 1866.

Bibliography

Cioc, Marc. *The Rhine: An Eco-Biography, 1815–2000.* Seattle: University of Washington Press, 2002.

Cohn, Hermann. *Untersuchungen der Augen von 10060 Schulkindern, nebst Vorschlägen zur Verbesserung der den Augen nachtheiligen Schuleinrichtungen: Eine ätiologische Studie.* Leipzig: Fleischer, 1867.

Colton, Timothy J. *Moscow: Governing the Socialist Metropolis.* Cambridge, MA: Belknap Press of Harvard University Press, 1995.

Cronon, William. *Nature's Metropolis: Chicago and the Great West.* New York: Norton, 1991.

Daly, Jonathan. "On the Significance of Emergency Legislation in Late Imperial Russia." *Slavic Review* 54, no. 3 (1995): 602–29.

Davis, John P. *Russia in the Time of Cholera: Disease under Romanovs and Soviets.* London: I. B. Tauris, 2018.

Davydov, A. N. *Moskva, vek XX. Istoricheskaia ekologiia.* Vol. 1. 1901–1917. Moscow: Mosgorarkhiv, 2000.

Davydov, A. N. "Vodosnabzhenie i kachestvo pit'ievoi vody v Moskve v XIX–nachale XX vekov." *Historia Provinciae*, 2, no. 1 (2018): 60–79.

Deiatel'nost' Moskovskogo gorodskogo upravleniia, 1913–1916. Moscow: Gorodskaia tipografiia, 1917.

Dement'iev, E. M. *Fabrika: chto ona daiet naseleniiu i chto ona u nego beret.* Moscow: Izd. T-va Sytina, 1897.

Denisov, L. A. *Stranitsy istorii sanitarnogo dela.* Moscow: Torius, 2014.

Dills, Randall. "The River Neva and the Imperial Façade: Culture and Environment in Nineteenth-Century Saint Petersburg Russia." PhD diss., University of Illinois at Urbana-Champaign, 2010.

Dobroslavin, A. P. *O sravnitel'noi stoimosti uluchshennoi pishchi arestantov s zatratami na ikh lechenie.* Saint Petersburg: Tipografiia Ia. Trei, 1884.

"Doklad N 41 ob ustroistve gorodskogo skotoprigonnogo dvora i boini." *Izvestiia Moskovskoi Gorodskoi Dumy* 3 (1885): 1–40.

"Doklad N 55 po voprosu o kanalizatsii Moskvy." *Izvestiia Moskovskoi Gorodskoi Dumy* 10 (1879): 1–39.

Donesenie Meditsinskomu sovetu Osoboi komissii po voprosu o trikhinakh v svinom miase. Saint Petersburg: Tipografiia MVD, 1876.

Drozdov, V. A. *Nekotorye dannye iz praktiki sanitarnoi tekhniki v dele okhrany rek ot zagriazneniia fabrichnymi vodami.* Moscow: Tipografii Mamontova, 1908.

Duisebayeva, Aibubi. "The Animal Face of Imperial Power: Kazakh Animal Husbandry and Tsarist Veterinary Services, 1868–1917." PhD diss., Al-Farabi Kazakh National University, 2023.

Dzhanshiev, G. A. *Epokha velikikh reform.* Saint Petersburg: Tipografiia Volfa, 1905.

Bibliography

Echenberg, Myron. *Plague Ports: The Global Urban Impact of Bubonic Plague, 1894–1901.* New York: New York University Press, 2007.

Egorysheva, I. V., E. V. Sherstneva, and S. G. Goncharova. *Meditsina gorodskikh obshchestvennykh samoupravlenii v Rossii.* Moscow: Shiko, 2017.

Eklof, Ben. *Russian Peasant Schools: Officialdom, Village Culture, and Popular Pedagogy, 1861–1914.* Berkeley: University of California Press, 1990.

Eklof, Ben. "Worlds in Conflict: Patriarchal Authority, Discipline and the Russian School, 1861–1914." *Slavic Review* 50, no. 4 (1991): 792–806.

Eklof, Ben, with Nadezhda Petersen. "Laska i Poriadok: The Daily Life of the Rural School in Late Imperial Russia." *Russian Review* 69 (January 2010): 7–29.

Elina, O. Iu. *Ot tsarskikh sadov do sovetskikh polei: Istoriia sel'skokhoziaistvennykh opytnykh uchrezhdenii, XVIII-20-e gody XX v.* Vols. 1–2. Moscow: RAN, 2008.

Engel, Barbara Alpern. *Between the Fields and the City: Women, Work and Family in Russia, 1861–1914.* Cambridge: Cambridge University Press, 1996.

Engel, Barbara Alpern. "The Woman's Side: Male Out-Migration and the Family Economy in Kostroma Province." *Slavic Review* 45, no. 2 (Summer 1986): 257–71.

Engelstein, Laura. *The Keys to Happiness: Sex and the Search for Modernity in Fin-de-Siecle Russia.* Ithaca, NY: Cornell University Press, 1992.

Engelstein, Laura. *Moscow, 1905: Working-Class Organization and Political Conflict.* Stanford, CA: Stanford University Press, 1982.

Entsiklopedicheskii slovar' Brokgauza-Efrona. 86 vols. Saint Petersburg: Brokgauz i Efron. 1890–1907.

Erickson, Arvel. "The Cattle Plague in England, 1865–1867." *Agricultural History* 35, no. 2 (1961): 93–104.

Erisman, F. F. *Kurs gigieny.* Vols. 1–2. Moscow: Tipografiia A. A. Kartseva, 1887–1888.

Erisman, F. F. *Kurs gigieny.* Vol. 1. Moscow: Tipolitografiia T-va I. N. Kushnerev i Ko, 1892.

Erisman, F. F., ed. *Novye kliniki i instituty (klinicheskii gorodok) Imperatorskogo Moskovskogo Universiteta na Devichiem pole.* Moscow: Tipolitografiia T-va I. N. Kushnerev i Ko, 1891.

Erisman, F. F. "Organizatsiia obshchestvennoi gigieny v Rossii." *Otechestvennyie zapiski* 6 (1876): 205–64.

Erisman, F. F. "Pishchevoe dovol'stvo rabochikh." In *Sbornik statisticheskikh svedenii po Moskovskoy gubernii. Otdel sanitarnoi statistiki. Obshchaia svodka po sanitarnym issledovaniiam fabrichnykh zavedenii Moskovskoi gubernii za 1879–1885 gg.*, vol. 4., part 2, 463–516. Moscow: I. N. Kushnerev, 1893.

Erisman, F. F. *Professional'naia gigiena umstvennogo i fizicheskogo truda.* Saint Petersburg: Tipografiia Stasiulevicha, 1877.

Bibliography

Erisman, F. F. "Sanitariia." In *Entsiklopedicheskii slovar' Brokgauza-Efrona*, vol. 56, 261–63. Saint Petersburg: Brokgauz i Efron, 1900.

Erisman, F. F. *Sanitarnoe issledovanie fabrichnykh zavedenii Klinskogo uiezda*. Moscow: Mosk. Gub. Zemstvo, 1891.

Erisman, F. F. *Sanitarnoe issledovanie fabrichnykh zavedenii Moskovskogo uiezda*. Moscow: Mosk. Gub. Zemstvo, 1882–1885.

Erisman, F. F., ed. *Sbornik rabot gigienicheskoi laboratorii Moskovskogo Universiteta*. Vols. 1–5. Moscow: Tipografiia Kartseva, 1886–1894.

Erisman, F. F. *Soobrazheniia po ustroistvu obraztsovoi klassnoi komnaty soglasno trebovaniiam sovremennoi gigieny*. Moscow: Pechatnia Iakovleva, 1888.

Erisman, F. F. *Soobrazheniia po voprosu o nailuchshem ustroistve klassnoi mebeli*. Moscow: Tipografiia Lissnera i Romana, 1894.

Erisman, F. F. *Vliianie shkol na proiskhozhdenie blizorukosti po nabliudeniiam nad uchashchimisia v uchebnykh zavedeniiakh v Sankt-Peterburge*. Saint Petersburg: Tipografiia Kotomina, 1870.

Erisman, F. F. "Znachenie bakteriologii dlia sovremennoi gigieny." In *Trudy vtorogo s'iezda russkikh vrachei v Moskve*, vol. 1, 18–38. Moscow: Pechatnia Iakovleva, 1887.

Evans, Richard J. *Death in Hamburg: Society and Politics in the Cholera Years, 1830–1910*. Oxford: Clarendon Press, 1987.

Fedorov, E. S. *Po povodu trudov Komissii pri Russkom Tekhnicheskom obshchestve po rassmotreniiu voprosov ob ochistke gorodov*. Kazan: Tipografiia Okr. Shtaba, 1885.

F[idler], V. *Moskva. Kratkie ocherki gorodskogo blagoustroistva*. Moscow: Tipografiia Blagushinoi, 1897.

Figes, Orlando. *The People's Tragedy: The Russian Revolution, 1891–1924*. New York: Penguin Books, 1998.

Filtzer, Donald. *The Hazards of Urban Life in Late Stalinist Russia: Health, Hygiene, and Living Standards, 1943–1953*. Cambridge: Cambridge University Press, 2010.

Finansy goroda Moskvy, 1863–1894. Moscow: Gorodskaia tipografiia, 1896.

Foucault, Michel. *Discipline and Punish: The Birth of the Prison*. New York: Vintage Books, 1995 [1975].

Frede, Victoria. *Doubt, Atheism and the Nineteenth-Century Russian Intelligentsia*. Madison: University of Wisconsin Press, 2011.

Freiberg, N. G. "Vrachebno-sanitarnoe zakonodatelstvo." In *Spravochnik po obshestvenno-sanitarnym i vrachebno-bytovym voprosam*, 181–210. Moscow: Tipographiia Richter, 1910.

Frenkel, Z. G. *Kholera i nashi goroda*. Moscow: Kushnerev, 1909.

Frenkel, Z. G. *Osnovy obshchego gorodskogo blagoustroistva*. Moscow: Izd. Glav. Upr. Kommunalnogo Khoziaistva NKVD, 1926.

Bibliography

Frieden, Nancy Mandelker. "The Russian Cholera Epidemic, 1892–1893, and Medical Professionalization." *Journal of Social History* 10, no. 4 (Summer 1977): 538–59.

Frieden, Nancy Mandelker. *Russian Physicians in an Era of Reform and Revolution, 1856–1905*. Princeton, NJ: Princeton University Press, 1981.

Fudge, Erica. "Milking Other Men's Beasts." *History and Theory* 52, no. 4 (December 2013): 13–28.

Gantner, Eszter, Heidi Hein-Kircher, and Oliver Hochadel, eds. *Interurban Knowledge Exchange in Southern and Eastern Europe, 1870–1950*. New York: Routledge, 2021.

Gantner, Eszter, Heidi Hein-Kircher, and Oliver Hochadel. "Introduction: Searching for Best Practices in Interurban Networks." In *Interurban Knowledge Exchange in Southern and Eastern Europe, 1870–1950*, edited by Eszter Gantner, Heidi Hein-Kircher, and Oliver Hochadel, 1–22. New York: Routledge, 2021.

Garcier, Romain. "The Placing of Matter: Industrial Water Pollution and the Construction of Social Order in Nineteenth-Century France." *Journal of Historical Geography* 36 (2010): 132–42.

Gerasimov, N. V. *Istoriia uluchsheniia Moskvy-reki*. Saint Petersburg: Tipografiia I. Goldberg, 1902.

Glavneishie predvaritelnye dannye perepisi goroda Moskvy 31 ianvaria 1902 g. Vol. 1. Moscow: Gorodskaia tipografiia, 1902.

Glavneishie predvaritelnye dannye perepisi goroda Moskvy 6 marta 1912 g. Moscow: S.n., 1913.

Gleason, William. "Public Health, Politics, and Cities in Late Imperial Russia." *Journal of Urban History* 16, no. 4 (1990): 341–65.

Goldberg, Carl Allan. "The Association of Industry and Trade, 1906–1917: The Successes and Failures of Russia's Organized Businessmen." PhD diss., University of Michigan, 1974.

Goldman, Wendy. "Women, Abortion, and the State, 1917–36." In *Russia's Women: Accommodation, Resistance, Transformation*, edited by Barbara Evans Clements, Barbara Alpern Engel, and Christine D. Worobec, 243–66. Berkeley: University of California Press, 1991.

Gorbunov, D. G. *Moskovskie gorodskie boini*. Moscow: Gorodskaia tipografiia, 1913.

Gornostaev, I. F. *Deti rabochikh i gorodskie popechitelstva o bednykh v Moskve*. Moscow: Kushnerev, 1900.

Gorshkov, Boris. *Russia's Factory Children: State, Society, and Law, 1800–1917*. Pittsburgh, PA: University of Pittsburgh Press, 2002.

Goubert, Jean-Pierre. *The Conquest of Water: The Advent of Health in the Industrial Age*. Princeton, NJ: Princeton University Press, 1989.

Grandin, Temple. "Making Slaughterhouses More Humane for Cattle, Pigs, and Sheep." *Annual Review of Animal Biosciences* 1 (January 2013): 491–512.

Bibliography

Guillaume, Louis. *Hygiène scolaire. Considérations sur l'état hygiénique des écoles publiques présentées aux autorités scolaires, aux institutions et aux parents.* Geneva, 1864.

Hachten, Elizabeth A. "In Service to Science and Society: Scientists and the Public in Late-Nineteenth-Century Russia." *Osiris* 17 (2002): 171–209.

Hachten, Elizabeth A. "Science in the Service of Society: Bacteriology, Hygiene and Medicine in Russia, 1855–1907." PhD diss., University of Wisconsin–Madison, 1991.

Hamlin, Christopher. *Public Health and Social Justice in the Age of Chadwick: Britain, 1800–1854.* Cambridge: Cambridge University Press, 1998.

Hard, Mikael, and Thomas J. Misa. "Modernizing European Cities: Technological Uniformity and Cultural Distinction." In *Urban Machinery: Inside Modern European Cities*, edited by Mikael Hard and Thomas J. Misa, 1–20. Cambridge, MA: MIT Press, 2008.

Hard, Mikael, and Thomas J. Misa, eds. *Urban Machinery: Inside Modern European Cities.* Cambridge, MA: MIT Press, 2008.

Hardy, Anne. *The Epidemic Streets: Infectious Diseases and the Rise of Preventive Medicine, 1856–1900.* Oxford: Oxford University Press, 1993.

Hearne, Siobhán. *Policing Prostitution: Regulating the Lower Classes in Late Imperial Russia.* Oxford: Oxford University Press, 2021.

Henze, Charlotte. *Disease, Health Care and Government in Late Imperial Russia: Life and Death on the Volga, 1823–1914.* New York: Routledge, 2011.

Hietala, Marjatta. *Services and Urbanization at the Turn of the Century: The Diffusion of Innovations.* Helsinki: Finnish Historical Society, 1987.

Hime, T. W. "Shall We Eat Tuberculous Meat?" *British Medical Journal* 1, no. 1528 (1890): 865–66.

Hofmann, Michèle. *Gesundheitswissen in der Schule: Schulhygiene in der deutschsprachigen Schweiz im 19. und 20. Jahrhundert.* Bielefeld: Transcript Verlag, 2016.

Hutchinson, John. "Tsarist Russia and the Bacteriological Revolution." *Journal of the History of Medicine and Allied Sciences* 40, no. 4 (1985): 420–39.

Hutchinson, John. *Politics and Public Health in Revolutionary Russia, 1890–1918.* Baltimore, MD: Johns Hopkins University Press, 1990.

Iablokov, N. V. *Prizrenie detei v vospitatel'nykh domakh.* Saint Petersburg: Gosudarstvennaia tipografiia, 1901.

Iavorskii, P. T. *Gorodskoi veterinarnyi nadzor v Moskve.* Moscow: Gorodskaia tipografiia, 1896.

Illiustrirovannyi putevoditel' po Moskve. Moscow: Dobrovolskii, 1911.

Issledovanie sovremennogo sostoianiia skotovodstva v Rossii. Vol. 1. Moscow: Topografiia Lavrova, 1884.

Istoriia Imperatorskoi Voenno-meditsinskoi (byvshei Voenno-khirurgicheskoi) Akademii za sto let, 1798–1898. Saint Petersburg: Tipografiia MVD, 1898.

Bibliography

Johnson, Robert E. "Peasant and Proletarian: Migration, Family Patterns and Regional Loyalties." In *The World of the Russian Peasant: Post-emancipation Culture and Society*, edited by Ben Eklof and Stephen Frank, 81–100. Boston: Unwin Hyman, 1990.

Jones, Ryan Tucker. *Empire of Extinction: Russians and the North Pacific's Strange Beasts of the Sea, 1747–1867*. Oxford: Oxford University Press, 2014.

Joyce, Patrick. *The Rule of Freedom: Liberalism and the Modern City*. London: Verso, 2003.

K. P. Pobedonostsev i ego korrespondenty. Moscow: Gosudarstvennoe izdatel'stvo, 1923.

Kablukov, K. A. *Studencheskii kvartirnyi vopros v Moskve*. Moscow: Obshchestvo vzaimopomoshchi studentov-iuristov, 1908.

Kamenetskaia, E. N. *Blagotvoritel'naia deiatel'nost' Moskovskogo gorodskogo obshchestvennogo uprvaleniia*. Moscow: Gorodskaia tipografiia, 1896.

Kamenetskaia, E. N. *Kanalizatsiia goroda Moskvy*. Moscow: Gorodskaia tipografiia, 1896.

Kanalizatsiia goroda Moksvy [Karty]. Moscow: S.n., 1912.

Kastal'skii, A. V., and V. F. Kozhinov. *Vodosnabzhenie i kanalizatsiia*. Moscow: Gos. Izdatel'stvo stroitel'noi literatury, 1941.

Kastal'skii, V. D. *Moskva v sovremennom sostoianii i chto ei predstoit sdelat' v otnoshenii blagoustroistva*. Moscow: Tipografiia Klein, 1883.

Kastal'skii, V. D. *O razdel'noi sisteme splavnoi kanalizatsii gorodov*. Moscow: Gorodskaia tipografiia, 1889.

Kelly, Catriona. *Children's World: Growing Up in Russia, 1890–1991*. New Haven, CT: Yale University Press, 2007.

Khodarkovsky, Michael. *Russia's Steppe Frontier: The Making of a Colonial Empire, 1500–1800*. Bloomington: Indiana University Press, 2002.

Kinnicutt, Leonard P., C.-E. A. Winslow, and R. Winthrop Pratt. *Sewage Disposal*. 2nd ed. New York: John Wiley and Sons, 1919.

Kishkin, N. "Zhilishchnyi vopros v Moskve i blizhaishie zadachi v ego razreshenii gorodskoi dumoi." *Gorodskoe delo* 5 (1913): 291–300 and 6 (1913): 351–60.

Kleptsov, N. Z. "O syrosti kamennykh zdanii i kolichestvennoe opredelenie ee." In *Trudy vtorogo s'iezda russkikh vrachei v Moskve*, vol. 1, 29–35. Moscow: Pechatnia Iakovleva, 1887.

Knight, Nathaniel. "Was the Intelligentsia Part of the Nation? Visions of Society in Post-emancipation Russia." *Kritika: Explorations in Russian and Eurasian History* 7, no. 4 (Fall 2006): 733–58.

Koch, Robert. "An Address on the Transference of Bovine Tuberculosis to Man." *British Medical Journal* 2, no. 2190 (December 1902): 1885–89.

Kolganova, E. V. "Zarozhdenie sistemy okhrany materinstva i mladenchestva v Rossii v kontse XIX–nachale XX vv." PhD diss., Moscow State University, 2012.

Bibliography

Kolokolov, M. M. *Vody Sankt-Peterburga, issledovannye kolichestvennym bakterioskopicheskim analizom*. Saint Petersburg: Tipografiia Evdokimova, 1886.

Kopsidis, Michael, Katja Bruisch, and Daniel W. Bromley. "Where Is the Backward Russian Peasant? Evidence against the Superiority of Private Farming, 1883–1913." *Journal of Peasant Studies* 42, no. 2 (2015): 425–47.

Korf, N. A. *Russkaia nachal'naia shkola*. 6th ed. Saint Petersburg: Tipografiia Hana, 1897.

Koroleva, N. G., A. P. Karelin, and L. F. Pisar'kova, eds. *Zemskoe samoupravlenie v Rossii, 1864–1918*. Moscow: Nauka, 2005.

Kotsonis, Yanni. *Making Peasants Backward: Agricultural Cooperatives and the Agrarian Question in Russia, 1861–1914*. Basingstoke, UK: Macmillan Press, 1999.

Kotsyn, M. B. *Opyt sistematicheskikh nabliudenii nad kolebaniem khimicheskogo i bakteriologicheskogo sostava vody Moskvy-reki za 1887–1888 g*. Moscow: Tipografiia Bonch-Bruevicha, 1889.

Koz'min, B. P. *Zubatov i ego korrespondenty: Sredi okhrannikov, zhandarmov i provokatorov*. Moscow and Leningrad: Gos. izdatel'stvo, 1928.

Kraikovski, Alexei, and Julia Lajus. "Living on the River over the Year: The Significance of the Neva to Imperial Saint Petersburg." In *Rivers Lost, Rivers Regained: Rethinking City-River Relations*, edited by Martin Knoll, Uwe Lübken, and Dieter Schott, 235–52. Pittsburgh, PA: University of Pittsburgh Press, 2017.

Krasnopevkov, A. "Obzor uchilishch dlia narodnogo obrazovaniia v Moskve." *Zhurnal Ministerstva Narodnogo Prosveshcheniia* 54, no. 4 (April 1871): 170–80.

"Kratkii otchet o deiatel'nosti gorodskikh ambulatorii za 5 let (1887–1891)." *Izvestiia Moskovskoi Gorodskoi Dumy* 3 (1892): 11–16.

Kriukov, N. A. *Miaso i miasnye zhivotnye*. Saint Petersburg: Kirshbaum, 1913.

Krug, Peter. "The Debate over the Delivery of Health Care in Rural Russia: The Moscow Zemstvo, 1864–1878." *Bulletin of the History of Medicine* 50, no. 2 (1976): 226–41.

Krylov, V. P. *K istorii trikhinoza v Rossii*. Moscow: Universitetskaia tipografiia, 1876.

Kupriianova, L. "Rabochii vopros v Rossii vo vtoroi polovine XIX–nachale XX vv." In *Istoriia predprinimatel'stva v Rossii*, vol. 2, edited by V. I. Bovykin, M. L. Glavlin, and M. L. Epifonova, 343–437. Moscow: ROSSPEN, 2000.

Kurkin, P. I. *Estestvennoie dvizheniie naseleniia g. Moskvy i Moskovskoi gubernii: statisticheskii obzor*. Moscow: Moszdravotdel, 1927.

Kurkin, P. I. "Sanitarnaia statistika." In *Spravochnik po obshchestvenno-sanitarnym i vrachebno-bytovym vorposam*, 89–127. Moscow: Tipografiia Richter, 1910.

Kuz'min, V. Iu. *Vlast', obshchestvo i zemskaia meditsina, 1864–1917*. Samara: Samarskii universitet, 2003.

La Berge, Ann. "Medicalization and Moralization: The Crèches in Nineteenth-Century Paris." *Journal of Social History* 25, no. 1 (1991): 65–87.

Bibliography

La Berge, Ann. *Mission and Method: The Early Nineteenth-Century French Public Health Movement.* Cambridge: Cambridge University Press, 2002.

LeBanc, Ronald. "Tolstoy's Way of No Flesh: Abstinence, Vegetarianism and Christian Physiology." In *Food in Russian History and Culture,* edited by Musya Glants and Joyce Toomre, 81–102. Bloomington: Indiana University Press, 1997.

Letnie kolonii moskovskikh gorodskikh nachalnykh uchilishch. Otchet, 1890–1915. Moscow: Tipografiia Mamontovoi, 1890–1916.

Levenson, D. M. *Gigiena shkol.* Odessa: Gorodskaia tipografiia, 1872.

Lindenmayer, Adele. *Poverty Is Not a Vice: Charity, Society, and the State in Imperial Russia.* Princeton, NJ: Princeton University Press, 1996.

Loskutova, M. V., and A. A. Fedotova. *Stanovlenie prikladnykh biologicheskikh issledovanii v Russii: Vzaimodeistvie nauki i praktiki v XIX–nachale XX vv.* Saint Petersburg: Nestro-Istoriia, 2014.

Maclachlan, Ian. "A Bloody Offal Nuisance: The Persistence of Private Slaughter-Houses in Nineteenth-Century London." *Urban History* 34, no. 2 (2007): 227–54.

Malinova-Tsiafeta, Olga. *Iz goroda na dachu: Sotsial'no-kul'turnye faktory osvoeniia dachnogo prostranstva vokrug Peterburga, 1860–1914.* Saint Petersburg: Izdatel'stvo Evropeiskogo Universiteta, 2013.

Martin, Alexander. *Enlightened Metropolis Constructing Imperial Moscow, 1762–1855.* Oxford: Oxford University Press, 2013.

Martin, Alexander. "Sewage and the City: Filth, Smell, and Representations of Urban Life in Moscow, 1770–1880." *Russian Review* 67, no. 2 (2008): 243–74.

Mazanik, Anna. "The City as a Transient Home: Residential Patterns of Moscow Workers around the Turn of the Twentieth Century." *Urban History* 40, no. 1 (2013): 51–70.

Mazanik, Anna. "The City of Men: Gender, Space and Working-Class Domesticity in Late-Imperial Moscow." In *Gendering Spaces in European Towns, 1500–1914,* edited by Elaine Chalus and Marjo Kaartinen, 114–31. New York: Routledge 2019.

Mazanik, Anna. "Industrial Waste, River Pollution and Water Politics in Central Russia, 1880–1917." *Water History* 10 (2018): 207–22.

Mazanik, Anna. "Learning from Smaller Cities: Moscow in the International Urban Networks, 1870–1910." In *Interurban Knowledge Exchange in Southern and Eastern Europe, 1870–1950,* edited by Eszter Gantner, Heidi Hein-Kircher, and Oliver Hochadel, 119–33. New York: Routledge, 2021.

Mazanik, Anna. "Managing the Catastrophe: Cholera, Urban Community and Health Politics in Imperial Moscow." In *Catastrophe, Gender and Urban Experience, 1648–1920,* edited by Deborah Simonton and Hannu Salmi, 198–213. New York: Routledge, 2017.

Bibliography

Mazanik, Anna. "Public Health across Species: Domestic Animals and Sanitary Reforms in Late Imperial Russia." In *Thinking Russia's History Environmentally*, edited by Catherine Evtukhov, Julia Lajus, and David Moon, 174–200. New York: Berghan Books, 2023.

Mazanik, Anna. "Sanitation, Urban Environment and the Politics of Public Health in Late Imperial Moscow." PhD diss., Central European University, 2015.

Mazanik, Anna. "School Doctors, Hygiene, and the Medicalization of Education in Imperial Moscow, 1889–1914." In *New Europe College Yearbook*, 105–29. Bucharest: New Europe College, 2018.

Mazanik, Anna. "'Shiny Shoes' for the City: The Public Abattoir and the Reform of Meat Supply in Imperial Moscow." *Urban History* 45, no. 2 (2018): 214–32.

Mazanik, Anna. "Vospriiatie nasiliia i zhestokosti na Moskovskikh gorodskikh boiniakh." In *Antropologiia Moskvy: Novoe znanie o gorode*, 26–33. Moscow: Muzei Moskvy, 2017.

Meckel, Richard. *Classrooms and Clinics: Urban Schools and the Protection and Promotion of Child Health, 1870–1930*. New Brunswick, NJ: Rutgers University Press, 2013.

Medovikov, P. S. *V chem dolzhna sostoiat' bor'ba s detskoi smertnostiiu*. Petrograd: Gosudarstvennaia tipografiia, 1916.

Melosi, Martin. *Garbage in the Cities: Refuse Reform and the Environment*. Pittsburgh, PA: University of Pittsburgh Press, 2004.

Melosi, Martin. *Precious Commodity: Providing Water for America's Cities*. Pittsburgh, PA: University of Pittsburgh Press, 2011.

Melosi, Martin. *The Sanitary City: Environmental Services in Urban America from Colonial Times to the Present*. Pittsburgh, PA: University of Pittsburgh Press, 2008.

Melosi, Martin. *The Sanitary City: Urban Infrastructure in America from Colonial Times to the Present*. Baltimore, MD: Johns Hopkins University Press, 2000.

Mezhduvedomstvennaia komissiia po peresmotru vrachebno-sanitarnogo zakonodatel'stva. Trudy Otdela Okhrany materinstva i bor'by s detskoi smertnost'iu. S.l.: S.n., 191-.

Mikhailov, N. *Pamiati professora F. F. Erismana*. Moscow: Tipo-litogragiia T-va I. I. Kushnerev i Ko, 1915.

Mikhailov, N. F. *Materialy k opredeleniiu fizicheskogo razvitiia i boleznennosti v sel'skikh shkolakh Ruzskogo uiezda Moskovskoi gubernii*. Moscow: Tipografiia Islenieva, 1887.

Mikhailov, N. F. *Obshchaia kharakteristika deiatel'nosti nashikh vospitatel'nykh domov*. Moscow: Pechatnia Iakovleva, 1887.

Miller, Aleksei, and Dmitry Chernyi, eds. *Goroda imperii v gody velikoi voiny i revolutsii*. Moscow: Nestor-Istoriia, 2017.

Miller, O. K. *K otsenke sposobov ochistki fabrichnykh stochnykh vod*. Moscow: Tipografiia Gerbek, 1892.

Bibliography

Mironov, Boris. *A Social History of Imperial Russia*. Boulder: Westview Press, 2000.

Moiseieva, T. A. *Zemskaia meditsina Simbirskoi gubernii*. Stavropol: Logos, 2019.

Moon, David. *The Plough That Broke the Steppe: Agriculture and Environment on Russia's Grasslands, 1700–1914*. Oxford: Oxford University Press, 2013.

Morris, Robert. "Governance: Two Centuries of Urban Growth." In *Urban Governance: Britain and beyond since 1750*, edited by Robert Morris and Richard H. Trainor, 1–14. Aldershot, UK: Ashgate, 2000.

Moskovskaia gorodskaia duma, 1913–1916. Moscow: Gorodskaia tipografiia, 1916.

Moskovskaia gorodskaia uprava. Obzor po gorodu Moskve za 1910 g. Moscow: Tipografiia Moskovskogo gradonachal'stva, 1911.

Moskovskie gorodskie nachal'nye uchilishcha. Statisticheskii otchet za 1901–1902 g. Moscow: S.n., 1903.

Moskva. Putevoditel'. Moscow: Kushnerev i K°, 1915.

Moskva. Vidy nekotorykh gorodskikh mestnostei, khramov, primechatelnykh zdanii i drugikh gorodskikh sooruzhenii. Moscow: Sherer, 1884–1891.

Moskva kak potrebitel'skii tsentr miasnykh produktov. Doklad Komissii boenskikh veterinarnykh vrachei Pervomu Mezhdunarodnomu Kongressu po kholodil'nomu delu v Parizhe v 1908 g. Moscow: Gorodskaia tipografiia, 1908.

Mosley, Stephen. *The Chimney of the World: A History of Smoke Pollution in Victorian and Edwardian Manchester*. New York: Routledge, 2008.

Nachal'nye uchilishcha, uchrezhdennye Moskovskoi Gorodskoi Dumoi. Moscow: Gorodskaia tipografiia, 1882.

Nagorskii, V. F. *O vliianii shkol na fizicheskoe razvitie detei*. Saint Petersburg: Tipografiia Ministerstva putei soobshcheniia, 1881.

Nagorskii, V. F. "Veterinarny nadzor na gorodskikh boiniakh g. Moskvy s ikh otkrytiia po 1 sentiabria." *Izvestiia Moskovskoi Gorodskoi Dumy* 9 (1888): 1–6.

Nardova, V. A. *Gorodskoe samoupravlenie v Rossii v 60-kh–nachale 90-kh godov XIX v.* Leningrad: Nauka, 1984.

Nardova, V. A. *Samoderzhavie i gorodskie dumy v Rossii v kontse XIX–nachale XX veka*. Saint Petersburg: Nauka, 1994.

Naumov, A. M. *O pitatel'nykh veshchestvakh i o vazhneishikh sposobakh ratsional'nogo ikh prigotovleniia, sberezheniia i otkrytiia v nikh primesei*. Saint Petersburg: Izdanie Torg. Doma S. Strugovshchikova, G. Pokhitonova, N. Vodova i K°, 1859.

Nelson, Amy, "The Body of the Beast: Animal Protection and Anticruelty Legislation in Imperial Russia." In *Other Animals: Beyond the Human in Russian Culture and History*, edited by Jane Costlow and Amy Nelson, 98–106. Pittsburgh, PA: University of Pittsburgh Press, 2010.

Nikitenko, V. P. *Detskaia smertnost' v Evropeiskoi Rossii za 1893–1896 god*. Saint Petersburg: Tovarishchestvo khudozhestvennoi pechati, 1901.

Nikitin, A. A. "Kanalizatsiia goroda Moskvy." In *Sovremennoe khoziaistvo goroda*

Bibliography

Moskvy, edited by I. A. Verner, 286–347. Moscow: Gorodskaia tipografiia, 1913.

Obertreis, Julia. *Imperial Desert Dreams: Cotton Growing and Irrigation in Central Asia, 1860–1991.* Göttingen: V&R Unipress, 2017.

"Obzor podgotovitelnykh rabot k sostavlennomu inzhenerom Gobrechtom proektu kanalizatsii Moskvy." *Izvestiia Moskovskoi Gorodskoi Dumy* 8 (1882): i–xlvi.

Ocherk deiatel'nosti Vserossiiskogo soiuza gorodov, 1914–1915. Moscow: S.n., 1916.

Ocherk vozniknoveniia i dvadtsatipiatiletnei deiate'nosti moskovskoi gorodskoi Detskoi bol'nitsy Sviatogo Vladimira, 1876–1900. Moscow: Gorodskaia tipografiia, 1901.

Okunev, N. P. *Dnevnik moskvicha.* Moscow: Voennoe izdatel'stvo, 1997.

O'Rourke, Share. *Warriors and Peasants: Don Cossacks in Late Imperial Russia.* Basingstoke, UK: Palgrave Macmillan, 2000.

"O sanitarnom sostoianii pomeshchenii, vnov' naniatykh dlia nachal'nykh gorodskikh uchilishch k nachalu 1902–1903 uchebnogo goda." *Izvestiia Moskovskoi Gorodskoi Dumy. Vrachebno-sanitarnyi otdel* (February 1903): 20–25.

Ostroglazov, V. M. *Smertnost' v Moskve.* Moscow: Universitetskaia Tipografiia, 1880–1893.

Ostrovskii, V. M. "Shkol'naia gigiena." In *Entsiklopedicheskii slovar' Brokgauza-Efrona,* vol. 39A, 633–43. Saint Petersburg: Brokgauz i Efron, 1903.

Otchet ekspertnoi komissii po voprosu ob ochistke stochnykh vod sakharnykh zavodov. Kiev: S.n., 1903.

Otchet Komissii po ochistke stochnykh vod, sostoiashchei pri Kanalizatsionnom otdele Moskovskoi gorodskoi upravy. Moscow: Gorodskaia tipografiia, 1913.

Otchet Meditsinskogo Departamenta za 1880 g. Saint Petersburg: MVD, 1882.

Otchet Moskovskoi gorodskoi upravy o merakh protiv rasprostraneniia aziatskoi kholery v 1892 g. Moscow: Gorodskaia tipografiia, 1893.

Otchet o deiatel'nosti Rossiiskogo Obshchestva Pokrovitel'stva Zhivotnym. Saint Petersburg: Tipografiia Soikina, 1867–1902.

Otchet o sostoianii narodnogo zdraviia i organizatsii vrachebnoi pomoshchi v Rossii. Saint Petersburg: MVD, 1877–1914.

"Otchet sanitarnogo vracha Arbatskoi chasti." *Izvestiiia Moskovskoi Gorodskoi Dumy* 16 (1878): 13–21.

"Otchet sanitarnogo vracha Meshchanckoi chasti." *Izvestiia Moskovskoi Gorodskoi Dumy* 17 (1878): 10–31.

"Otchet sanitarnogo vracha Prechistenskoi chasti." *Izvestiia Moskovskoi Gorodskoi Dumy* 14 (1878): 44–55.

Otchet veterinarnogo otdeleniia MVD. Saint Petersburg: MVD, 1885–1916.

Otchet vremennogo komiteta po izyskaniiu mer k okhrane vodoemov Moskovskogo promyshlennogo raiona ot zagriazneniia stochnymi vodami i otbrosami fabrik i zavodov. Moscow: S.n., 1913–1917.

Bibliography

Otchety moskovskikh gorodskikh sanitarnykh vrachei. Moscow: S.n., 1888–1894.

"Otnoshenie veterinarnogo vracha K.Z. Kleptsova v Komissiiu Obshchestvennogo Zdraviia." *Izvestiia Moskovskoi Gorodskoi Dumy* 5 (1892): 15–22.

Otter, Chris. "Civilizing Slaughter: The Development of the British Public Abattoir, 1850–1910." *Food and History* 3, no. 2 (2005): 29–51.

Otter, Chris. "Locating Materiality in Urban History." In *Material Powers: Cultural Studies, History and the Material Turn*, edited by Patrick Joyce and Tony Bennett, 38–59. London: Routledge, 2010.

Owen, Thomas. *Capitalism and Politics in Russia: A Social History of the Moscow Merchants, 1855–1905.* Cambridge: Cambridge University Press, 1981.

Owen, Thomas. "Doing Business in Merchant Moscow." In *Merchant Moscow: Images of Russia's Vanished Bourgeoisie*, edited by James West and Iurii Petrov, 29–36. Princeton, NJ: Princeton University Press, 1998.

"O zavtrakakh v nachal'nykh uchilishchakh goroda Moskvy." *Izvestiia Moskovskoi Gorodskoi Dumy. Vrachebno-sanitarnyi otdel* (August 1904): 37–44.

Ozerova, N. A. "Istoriia izucheniia gidrograficheskoi seti basseina r. Moskvy." PhD diss., Russian Academy of Sciences, 2010.

Pavlov, I. *Narodnaia shkola. Opyt razrabotki voprosa o narodnoi shkole so storony tekhnicheskoi, gigienicheskoi i ekonomicheskoi.* St. Petersburg: Tipografiia Balasheva, 1886.

Pavlova, S. K. *Sbornik podvizhnykh igr na otkrytom vozdukhe v shkole.* Moscow: Tipolitografiia Snegirevoi, 1896.

Pearson, Susan. "Speaking Bodies, Speaking Minds: Animals, Language, History." *History and Theory*, Theme issue 52 (December 2013): 91–108.

Pechorkin, E. F. "Ambulatoriia v ee nastoiashchem i blizhaishem budushchem." *Obshchestvennyi vrach* 6 (1912): 753–65.

Pervaia Vseobshchaia perepis' naseleniia Rossiiskoi imperii 1897 g. Vol. 2. Saint Petersburg: MVD, 1904.

Peterson, Maya. *Pipe Dreams: Water and Empire in Central Asia's Aral Sea Basin.* Cambridge: Cambridge University Press, 2019.

Petropavlovskii, N. P. *K voprosu o rasprostranenii trikhin sredi zhivotnykh goroda Khar'kova.* Saint Petersburg: Tipografiia MVD, 1899.

Petrov, Iu. A. *Moskovskaia burzhuaziia v nachale XX veka.* Moscow: Mosgorarkhiv, 2002.

Petrov, Iu. A. "Moskovskii delovoi mir na rubezhe XIX–XX vekov." In *Kupecheskaia Moskva: Obrazy ushedshei rossiiskoi burzhuazii*, edited by Iu. Petrov and J. West, 5–20. Moscow: ROSSPEN, 2004.

Petunnikov, A. "Gidrograficheskii ocherk Moskvy." *Izvestiia Moskovskoi Gorodskoi Dumy* 7 (1882): 70–74.

Petunnikov, A. "Materialy dlia izucheniia Moskvy v sovremennom eia sostoianii." *Izvestiia Moskovskoi Gorodskoi Dumy* 1 (1882): 46–54.

Bibliography

Petunnikov, A. "Moskva i eia budushchnost'." *Izvestiia Moskovskoi Gorodskoi Dumy* 1 (1881): 8–15.

Petunnikov, A. "Sostav i svoistva Moskovskikh vod." *Izvestiia Moskovskoi Gorodskoi Dumy* 3 (1879): 15–31.

Pirogovskaia, M. A. *Miasmy, simptomy, uliki: Zapakhi mezhdu meditsinoi i moraliu v russkoi kul'ture vtoroi poloviny XIX veka*. Saint Petersburg: Evropeiskii Universitet, 2018.

Pisar'kova, L. F. *Gorodskie reformy v Rossii i Moskovskaia Duma*. Moscow: Novyi Khronograf, 2010.

Poderni, S. A. *Tekhnicheskoe opisanie Moskovskikh tsentral'nykh gorodskikh boen*. Moscow: Gorodskaia tipografiia, 1896.

Poiasnitel'naia zapiska k proetku kanalizatsii goroda Moskvy. Moscow: S.n., 1890.

Poiasnitel'naia zapiska po shliuzovaniiu Moskvy-reki. Moscow: S.n., 1882.

Pokrovskaia, M. I. *Sanitarnyi nadzor nad zhilishchami i sanitarnaia organizatsioa v razlichnykh gosudarstvakh*. Saint Petersburg: Tipografiia Soikina, 1897.

Poleshchuk, K. K. "'Nevozmozhnaia v Moskve dolzhnost' gorodskogo golovy': Nikolai Alekseev i koronnaia administratsiia." *Rodina* 12 (2012): 106–8.

Poleshchuk, K. K. "Deiatel'nost' N. A. Alekseeva na postu Moskovskogo gorodskogo golovy, 1885–1893." PhD diss., Moscow State University, 2017.

P[olferov], Ia. "Chuma rogatogo skota." In *Entsiklopedicheskii slovar' Brokgauza i Efrona*, vol. 86, 52–54. Saint Petersburg: Brokgauz i Efron, 1903.

Popov, M. A. *Domovye i dvorovye stoki*. Saint Petersburg: Tipografiia Goppe, 1885.

Popov, M. A. "Kanalizatsiia goroda Moskvy: po proektu inzhener-gidrotekhnika M.A. Popova." *Izvestiia Moskovskoi Gorodskoi Dumy* 8 (1880): 5–32.

Pravilova, Ekaterina. *A Public Empire: Property and the Quest for the Common Good in Imperial Russia*. Princeton, NJ: Princeton University Press, 2014.

Pronshtein, A. P., ed. *Don i stepnoe Predkavkazie. Zaselenie i khoziaistvo*. Rostov-on-Don: Izdatel'stvo Rostovskogo universiteta, 1977.

Ransel, David. *Mothers of Misery: Child Abandonment in Russia*. Princeton, NJ: Princeton University Press, 1988.

Ratmanov, P. E. *Sovetskoe zdravookhranenie na mezhdunarodnoi arene v 1920-kh–1940-kh gg.: Mezhdu "miagkoi siloi" i propagandoi*. Khabarovsk: Izdatel'stvo DVGMU, 2021.

Rieber, Alfred J. *Merchants and Entrepreneurs in Imperial Russia*. Chapel Hill: University of North Carolina Press, 1982.

Robichaud, Andrew. *Animal City: The Domestication of America*. Cambridge, MA: Harvard University Press, 2019.

Rosenthal, Leslie. *The River Pollution Dilemma in Victorian England: Nuisance Law versus Economic Efficiency*. Farnham, UK: Ashgate Publishing, 2014.

Ruble, Blair. *Second Metropolis: Pragmatic Pluralism in Gilded Age Chicago, Silver Age Moscow, and Meiji Osaka*. Cambridge: Cambridge University Press, 2001.

Bibliography

Rudnev, M. M. *O trikhinakh v Rossii: Nereshennye vorposy v istorii trikhinnoi bolezni.* Saint Petersburg: Tipografiia Ia. Treia, 1866.

Sahadeo, Jeff. "Epidemic and Empire: Ethnicity, Class, and 'Civilization' in the 1892 Tashkent Cholera Riot." *Slavic Review* 64, no. 1 (2005): 117–39.

Samojlik Tomasz, Anastasia Fedotova, Piotr Daszkiewicz, and Ian D. Rotherham. *Białowieża Primeval Forest: Nature and Culture in the Nineteenth Century.* Cham: Springer, 2020.

Sanitarnye usloviia koechno-komorochnykh kvartir v rabochikh raionakh Moskvy. Moscow: Kushnerev, 1915.

"Sanitarnyi otchet po Iauzskoi chasti." *Izvestiia Moskovskoi Gorodskoi Dumy* 15 (1878): 34–44.

Saunier, Pierre-Yves, and Shane Ewan, eds. *Another Global City: Historical Explorations into the Transnational Municipal Moment, 1850–2000.* New York: Palgrave Macmillan, 2008.

Sbornik dekretov i postanovlenii po narodnomu khoziaistvu: izdanie Vysshego Soveta Narodnogo Khoziaistva. Vol. 2. Moscow: VSNKh, 1920.

Sbornik obiazatel'nykh dlia zhitelei g. Moskvy i chastiiu drugikh gorodov postanovlenii. Moscow: Pechatnia Iakovleva, 1879.

Sbornik obiazatel'nykh dlia zhitelei goroda Moskvy postanovlenii Moskovskoi Gorodskoi Dumy. Moscow: Gorodskaia tipografiia, 1896.

Sbornik postanovlenii Moskovskogo gubernskogo zemskogo sobraniia s 1865 po 1897 god Moscow: Pechatnia Iakovleva, 1901.

Schneider, Daniel. *Hybrid Nature: Sewage Treatment and Contradictions of the Industrial Ecosystem.* Cambridge, MA: MIT Press, 2011.

Shaw, David Gary. "A Way with Animals." *History and Theory* 52, no. 4 (December 2013): 1–12.

Sherstneva, E. V. "Moskovskoe gorodskoe samoupravlenie v bor'be s epidemiiami." *Gigiena i Sanitaria* 6 (2021): 647–52.

Shchepkin, M. P. *Obshchestvennoe samoupravlenie v Moskve. Proekt gorodovogo polozheniia.* Moscow: Levenson, 1906.

Sheail, John. "Town Wastes, Agricultural Sustainability and Victorian Sewage." *Urban History* 23, no. 2 (1996), 189–210.

Shevyrev, Alexander. "The Axis Petersburg–Moscow: Outward and Inward Russian Capitals." *Journal of Urban History* 30, no. 1 (2003): 70–84.

The Shone Hydro-Pneumatic System of Sewerage. Liverpool: Rockliff Brothers, 1885.

Smertnost' naseleniia goroda Moskvy 1872–1889. Moscow: Gorodskaia tipografiia, 1891.

Smith, Alison K. "Public Works in an Autocratic State: Water Supplies in an Imperial Russian Town." *Environment and History* 11, no. 3 (2005): 319–42.

Smolenskii, P. O. *Boini i skotoprigonnye dvory.* Saint Petersburg: Tipografiia zhurnala "Stroitel," 1902.

Bibliography

Snowden, Frank. *Naples in the Time of Cholera, 1884–1911.* Cambridge: Cambridge University Press, 2011.

Sobranie uzakonenii i rasporiazhenii pravitel'stva za 1912 god. Saint Petersburg: Tipografiia Pravitelstvuiushchego Senata, 1912, part 1.

Sokolov, A. D. *Rezultaty analizov vody reki Yauzy. Sutochnyie kolebaniia kisloroda v vode: Izvlecheniia iz raboty, udost. Med. fak. Mosk. un-ta zolotoi medali.* Saint Petersburg: Tipografiia Doma prizreniia maloletnikh bednykh, 1892.

Sokolov, D. A., and V. I. Grebenshchikov. *Smertnost' v Rossii i bor'ba s neiu.* Saint Petersburg: Tipografiia Stasiulevicha, 1901.

Solomon, Susan Gross, and John F. Hutchinson, eds. *Health and Society in Revolutionary Russia.* Bloomington: Indiana University Press, 1990.

Sputnik moskvicha: Moskva i ee okrestnosti. Moscow: Kushnerev i K°, 1894.

Stanchevici, Dmitri. *Stalinist Genetics: The Constitutional Rhetoric of T. D. Lysenko.* New York: Routledge, 2017.

Starks, Tricia. *The Body Soviet: Propaganda, Hygiene, and the Revolutionary State.* Madison: University of Wisconsin Press, 2008.

Statisticheskii atlas goroda Moskvy. Moscow: Gorodskaia tipografiia, 1890.

Statisticheskii atlas goroda Moskvy. Moscow: Gorodskaia tipografiia, 1911.

Statisticheskii ezhegodnik g. Moskvy. Moscow: Izdatel'stvo moskovskoi gorodskoi dumy, 1908–1916.

Statisticheskii ezhegodnik g. Moskvy i Moskovskoi gubernii. Vol 2. Statisticheskie dannye po g. Moskve za 1914–1925 gg. Moscow: S.n., 1927.

Statisticheskii ezhegodnik Rossii za 1915 g. Petrograd: MVD, 1916.

Strashun, I. D. *Russkaia obshchestvennaia meditsina v period mezhdu dvumia revolutsiiami 1907–1917.* Moscow: Meditsina, 1964.

Strobel, Angelika. *Die "Gesundung Russlands": Hygiene und imperial Verwaltungspraxis um 1900.* Bielefeld: Transcript Verlag, 2022.

Sunier, Pierre-Yves, and Shane Ewen, eds. *Another Global City: Historical Explorations into the Transnational Municipal Moment, 1850–2000.* New York: Palgrave Macmillan, 2008.

Sutugin, V. V. *Kratkii ocherk ustroistva Moskovsokgo rodovspomogatel'nogo zavedeniia i Meditsinskii otchet za 1888 g.* Moscow: Gorodskaia tipografiia, 1889.

Sviatlovskii, V. *O fabrichnykh otbrosakh i stochnykh vodakh promyshlennykh zavedenii.* Chernigov: Tipografiia Gubernskogo Pravleniia, 1889.

Swabe, Joanna. *Animals, Disease and Human Society: Human-Animal Relations and the Rise of Veterinary Medicine.* London: Routledge, 1999.

Sytin, P. V. *Kommunal'noe khoziaistvo (blagoustroistvo) Moskvy.* Moscow: Novaia Moskva, 1926.

Tablitsy o dvizhenii naseleniia v g. Moskve. Moscow: Gorodskaia tipografiia, 1889–1892.

Bibliography

Tarr, Joel. "From City to Farm: Urban Wastes and the American Farmer." *Agricultural History* 49, no. 4 (October 1975), 598–612.

Tarr, Joel A. "The Separate vs. Combined Sewer Problem: A Case Study in Urban Technology Design Choice." *Journal of Urban History* 5, no. 3 (1979): 313–19.

Tatarskii, V. "Veterinarnye instituty i shkoly." In *Entsiklopedicheskii slovar' Brokgauza-Efrona*, vol. 6, 132–33. Saint Petersburg: Brokgauz i Efron, 1892.

Thurston, Robert W. *Liberal City, Conservative State: Moscow and Russia's Urban Crisis, 1906–1914.* New York: Oxford University Press, 1987.

Tikhomirov, V. A. *O legchaishem sposobe otkrytiia trikhin v podozritel'nom miase.* Moscow: Universitetskaia tipografiia, 1875.

Todes, Daniel. *Ivan Pavlov: Russian Life in Science.* Oxford: Oxford University Press, 2014.

Tretiak, A. S. *Stanovlenie i razvitie zemskoi meditsiny i veterinarii v Tsentral'nom Chernozem'e: 1865–1914.* Kursk: Universitetskaia kniga, 2018.

Trudy III (Stroitel'nogo) otdela Imperatorskogo Russkogo tekhnicheskogo obshchestva, 1880–1884. Saint Petersburg: Tipografiia Panteleevykh, 1884.

Trudy Komissii dlia sostavleniia proekta polozheniia ob ustroistve i soderzhanii promyshlennykh zavedenii i nadzore za proizvodstvom na nikh rabot. Vol. 2. Saint Petersburg: S.n., 1896.

Trudy Moskovskogo gorodskogo statisticheskogo otdela. Vol. 2 Promyshlennye i torgovye zavedeniia Moskvy za 1879 god. Moscow: Moskovskaia gorodskaia tipografiia, 1882.

Turner, James. *Reckoning with the Beast: Animals, Pain, and Humanity in the Victorian Mind.* Baltimore, MD: Johns Hopkins University Press, 1980.

Uekoetter, Frank. "City Meets Country: Recycling Ideas and Realities on German Sewage farms." *Journal for the History of Environment and Society* 1 (2016): 89–107.

Uspenskii, V. P. *Moskva. Kratkii ocherk razvitiia i sovremennogo sostoianiia gorodskoi vrachebno-sanitarnoi organizatsii.* Moscow: Gorodskaia tipografiia, 1911.

Ustav Rossiiskogo Obshchestva pokrovitel'stva zhivotnym. Saint Petersburg: Tipografiia Retgera i Shneidera, 1865.

Ustav Rossiiskogo Obshchestva pokrovitel'stva zhivotnym. Saint Petersburg: Tipografiia Suvorina, 1888.

Vasilievskii, N. P. *Gigiena i sanitariia v primenenii k zemskim shkolam Khersonskoi gubernii.* Kherson: Khersonskaia gebernskaia zemskaia uprava, 1896.

Velychenko, Stephen. "Chislennost' biurokratii i armii v Rossiiskoi imperii v sravnitel'noi perspektive." In *Rossiiskaia imperiia v zarubezhnoi istoriografii*, edited by P. Werth, P. S. Kabytov, and A. I. Miller, 83–114. Moscow: Novoe Izdatel'stvo, 2005.

Verner, I. A., ed. *Sovremennoe khoziaistvo goroda Moskvy.* Moscow: Gorodskaia tipografiia, 1913.

Bibliography

Verner, I. A. *Zhilishcha bedneishego naseleniia g. Moskvy.* Moscow: Gorodskaia tipografiia, 1902.

Verner, K. A. "Moskovskii skotnyi i miasnoi rynok." *Izvestiia Moskovskoi Gorodskoi Dumy* 5 (1887): 24–56.

Veselovskii, B. B. *Istoriia zemstva za sorok let.* Vols.1–4. Saint Petersburg: Tipografiia Popovoi, 1909.

Veterinarnyi nadzor Moskovskikh Gorodskikh Boen. Moscow: Gorodskaia tipografiia, 1896.

Vialles, Noelie. *Animal to Edible.* Cambridge: Cambridge University Press, 1994.

Vigdorchik, N. A. "Voprosy narodnogo zdraviia i vrachebnogo byta. 1908 god." *Obshchestvennyi vrach* 1 (1909): 11–17.

Vil'iams, V. R. "Deiatel'nost' polei orosheniia Moskovskoi gorodskoi upravy v 1900 g." In *V. R. Vil'iams, Sobranie sochinenii*, vol. 2, 167–73. Moscow: Selkhozgiz, 1948.

Vil'iams, V. R. "Obshchie osnovaniia obezvrezhivaniia nechistot." In *Kanalizatsiia goroda Moskvy*, 18–27. Moscow: Gorodskaia tipografiia, 1901.

Vinogradov, Andrei. "Kazan' Citizens against Air Pollution: The Case of the Ushkov & Co. Chemical Factory (1893–1917)." In *Thinking Russia's History Environmentally*, edited by Catherine Evtuhov, Julia Lajus, and David Moon, 98–122. Oxford: Berghahn Books, 2023.

Virchow, Rudolph. *Ueber gewisse die Gesundheit benachtheiligende Einflüsse der Schulen: ein Bericht.* Berlin: Reimer, 1869.

Virenius, A. S. *Organizatsiia sanitarnoi chasti v uchebnykh zavedeniiakh.* Saint Petersburg: Tipografiia Balasheva, 1886.

Virenius, A. S. *Shkol'nye stoly i skam'i, ikh ustroistvo i raspredelenie v uchebnykh zavedeniiakh.* Saint Petersburg: Tipografiia Stasiulevicha, 1886.

Virkhov, Rudolf. *Izlozhenie ucheniia o trikhinakh: S ukazaniem na preduprediteľnyie mery etoi bolezni.* Saint Petersburg: Tipografiia Baksta, 1864.

Virkhov, Rudolf. "O vrednykh vliianiiakh shkoly na zdorovie." *Zhurnal Ministerstva Narodnogo Prosveshcheniia* 146, part 3 (December 1869): 266–87.

Vishnevskii, A. N. *Kniaz' Vladimir Andreyevich Dolgorukov, byvshii moskovskii gubernator.* Moscow: S.n., 1910.

Vodosnabzhenie goroda Moskvy v 1779–1902 gg. Moscow: Gorodskaia tipografiia, 1902.

Vremennik Tsentral'nogo Statisticheskogo Komiteta MVD. Vypusk 50. Svedeniia o kolichestve skota v 1900 g. Saint Petersburg: MVD, 1901.

Walker, Lisa Kay. "The Pen and the Test-Tube Revisited: Generational Identity, Scientific Innovation, and Russia's Reception of Bacteriology, 1890–1914." *Soviet and Post-Soviet Review* 32, no. 2–3 (2005): 269–91.

Walker, Lisa Kay. "Public Health, Hygiene and the Rise of Preventive Medicine in Russia, 1874–1912." PhD diss., University of California, Berkeley, 2003.

Bibliography

Watts, Sydney. "Liberty, Equality, and the Public Good: Parisian Butchers and Their Right to the Marketplace during the French Revolution." *Food and History* 3, no. 2 (2005): 105–17.

White, Elizabeth. *A Modern History of Russian Childhood from the Late Imperial Period to the Collapse of the Soviet Union.* London: Bloomsbury Academic, 2020.

Wilkinson, Lise. *Animals and Disease: An Introduction to the History of Comparative Medicine.* Cambridge: Cambridge University Press, 1992.

Wilkinson, Lise. "Glanders: Medicine and Veterinary Medicine in Common Pursuit of a Contagious Disease." *Medical History* 25, no. 4 (1981): 363–84.

Williams, Christopher. *Health and Welfare in Saint Petersburg, 1900–1941: Protecting the Collective.* New York: Routledge, 2018.

Winiwarter, Verena. "Where Did All the Waters Go? The Introduction of Sewage in Urban Settlements." In *Environmental Problems in European Cities in the 19th and 20th Centuries,* edited by Christoph Bernhardt, 105–19. Munster: Waxmann, 2004.

Winslow, C.-E. A. "Public Health Administration in Russia in 1917." *Public Health Reports* 32, no. 52 (December 28, 1917): 2191–219.

Vitte, S. Iu. *Vospominaniia.* Moscow: Izdatel'stvo sotsial'no-ekonomicheskoi literatury, 1960.

Worboys, Michael. *Spreading Germs: Disease Theories and Medical Practice in Britain, 1865–1900.* Cambridge: Cambridge University Press, 2000.

Worboys, Michael. "Was There a Bacteriological Revolution in Late Nineteenth-Century Medicine?" *Studies in History and Philosophy of Science, Part C: Studies in History and Philosophy of Biological and Biomedical Sciences* 38, no. 1 (2007): 20–42.

World Health Organization (WHO). *Housing and Health Guidelines.* Geneva: World Health Organization, 2018. https://www.ncbi.nlm.nih.gov/books/NBK535289/.

Young Lee, Paula, ed. *Meat, Modernity and the Rise of the Slaughterhouse.* Durham: University of New Hampshire Press, 2008.

"Zaiavlenie glasnogo A.D. Lopasheva." *Izvestiia Moskovskoi Gorodskoi Dumy* 4 (1879): 1–3.

"Zapiska o rabotakh Komissii po ustroistvu boen do 1885 g." *Izvestiia Moskovskoi Gorodskoi Dumy* 3 (1885): 1–8.

Zatravkin, Sergei, and Elena Vishlenkova. *"Kluby" i "getto" sovetskogo zdravookhraneniia.* Moscow: Shiko, 2022.

Zavolzhskaia, Iu. I. *Shkol'naia gigiena.* Saint Petersburg: Izdanie zhurnala "Sovremennaia Meditsina i Gigiena," 1898.

Zelenin, N. V. "Moskovskie gorodskie boini." In *Sovremennoe khoziaistvo goroda Moskvy,* edited by I. A. Verner, 466–521. Moscow: Gorodskaia tipografiia, 1913.

Bibliography

Zeyfman, P. T. *Trikhiny i trikhinnaia bolezn'*. Saint Petersburg: Tipografiia Ia. Treia, 1877.

Zhbankov, D. N. "Kratkie svedeniia o vozniknovenii i deiatel'nosti obshchestven-no-sanitarnykh uchrezhdenii v zemskoi Rossii." In *Spravochnik po obshchestven-no-sanitarnym i vrachebno-bytovym voprosam*, 38–88. Moscow: Tipographiia Richter, 1910.

Zhbankov, D. N. *Sbornik po gorodskomu vrachebno-sanitarnomu delu v Rossii*. Moscow: Tipografiia Richter, 1915.

Zhenkov, M. N., ed. *Ocherk deiatel'nosti Vserossiiskogo soiuza gorodov, 1914–1915*. Moscow: S.n., 1916.

Zhurnaly Komissii po nadzoru za ustroistvom novogo vodoprovoda i kanalizatsii v Moskve. Moscow: Gorodskaia tipografiia, 1891–1915.

Zimin, N. "K voprosu o vodosnabzhenii Moskvy." *Izvestiia Moskovskoi Gorodskoi Dumy* 7 (1879): 26–33.

Zviaginskii, Ia. Ia. *Domovaia kanalizatsiia, ee ustroistvo i ekspluatatsiia*. Moscow: Kushnerev, 1912.

Zviagintsev, E. A., M. N. Kovalenskii, M. S. Sergeev, and K. V. Sivkov, eds. *Moskva*. Moscow: I. N. Kushnerev, 1915.

INDEX

Note: Page numbers in *italics* indicate figures.

abattoir: as a profitable enterprise, 49, 121; built environment, 9, *123*, 124–25, 141–43; international prototypes, 115–16, 125; meat control, 125–30; public image, 116, 130–31; treatment of animals during slaughter, 134–39; working conditions, 139–45
abolition of serfdom, 16–17, 58, 79, 86
Agricultural Academy, 62, 80, 125
Agricultural Institute. *See* Agricultural Academy
Albrecht, Emmanuil, 169
Alekseev, Nikolai, 31–37, 40–43, 49, 169, 170, 185
Alexander II (1855–1881), 44, 87, 133
Alexander III (1881–1894), 33–34, 45, 82, 133
ambulatoriia. *See* outpatient clinics
Andreeva, Olga, 186
anemia, 176, 197
animal disease, 41, 108–14, 126–29
animal welfare, 132–39, 145
anthrax, 111–12, 126, 130, 224n34
Antwerp, 125
auxiliary classes for children with special needs, 193–94, 199

backwardness: of Moscow vs. Western cities, 5–7, 30, 35, 60, 71, 122; of peasants, 81–82, 84
bacteriological stations, 22–23, 212n18
bacteriology, 8, 22–26, 60, 67–68, 95, 112–13, 192
Belgium, 124. *See also* Brussels
Berlin, 17, 54–55, 62, 78, 122, 155, 159
birthing houses, 161, *162*, 163–64, 203
Black Sea, 9, 102, 104, 106
bovine pleuropneumonia, 111–12, 128
bovine tuberculosis, 126, 128–29
breastfeeding, 151–52, 164, 167, 171
Britain, 8, 78, 86, 89, 108, 124, 130–33
Brussels, 122
Bubnov, Sergei, 149
Building Statute, 86

Caspian Sea, 9, 102, 104, 106
cats, 112–13
cattle plague. *See* rinderpest
cattle: at the abattoir, 122–31; cattle drives, 9, 101–7; city cattle in Moscow, 112, 116–17; in the Moscow Province, 79–80, 83; railway transportation of cattle, 109–10

257

Index

cattle-driving trails, 105–6
Caucasus, 104–6
Central Asia, 9, 54, 104–6
cesspools, 36, 53, 59–60, 74–78, 85–87
Chadwick, Edwin, 64
Chicherin, Boris, 33–34
child abandonment, 152–57, 161
childbirth, 152–53, 161, 163, 171
childcare, 25, 150–51, 161–67
child-centered classroom, 186, 231n38
cholera: 1892 epidemic, 44–46, 89–90, 124, 128; and industrial pollution, 89–91, 96, 124, 128, 168; and the sewage system, 60–61, 67; in the twentieth century, 168, 208, 232n12
community medicine: and livestock, 111–14; and schooling, 174, 192, 200; development of, 26–27; rural, 7–8, 20–21, 28, 152, 204; transnational connections, 43; urban, 7–8, 14, 32–33, 41, 45–46, 205
compensation, 39, 79, 109, 112, 127–28, 140–44
compulsory killing of animals, 109–13, 128
concession, 64, 67–68, 121
confiscation, 127
consultations for parents, 164, 171
corporal punishment, 178, 192
Cossacks, 104, 106
cot-and-corner apartments, *158*, 159–60, 168–69
cowboys. *See* livestock trade
cruelty to animals, 133–38

Decree on Measures for the Preservation of State Order and Social Stability, 45
dentistry, 49, 191

diarrheal diseases in children, 151, 157, 160
disinfection, 4, 39, 169, 184, 192, 203, 207
Dmitrii Konstantinovich, Grand Duke, 133
Dnieper, 104, 106
dogs, 112–13
Dolgorukov, Vladimir, 33–34, 43
Don, 104, 106–7
drinking water, 25, 57, 59, 62, 90
dystentery, 160, 208

education for girls, 174, 181–83, 191, 193
epizootics. *See* animal disease
Erismann, Friedrich: career, 20–21, 28, 40–41, 43, 149; on hygiene and bacteriology 21–25; on school hygiene 175–80, 185; on the sewage system 66; on water pollution 59–60
Europeanness, 7, 60–69, 122–24, 209

factories: and children, 176, 183; and pollution, 58, 74, 85–98; and women, 153, 165; at the abattoir 119, 142; sanitary conditions 28, 39
factory inspection, 86, 88, 95
famine, 44
feedlots, 105, 107, 113
filtration fields, 61, 66, *77*, 78–84, *81*, 92, 119–25
fines, 45, 87, 93–94, 97, 133
First World War, 49, 54, 89, 96, 205–8
fodder, 79–80, 103, 105, 107, 113
food control, 25, 39–40, 207. *See also* abattoir
foot-and-mouth disease, 112
France, 8, 24, 89, 124, 130–31, 151. *See also* Paris

Index

gastroenteritis. *See* diarrheal diseases in children
Geneva, 122
germ theory of disease. *See* bacteriology
Germany, 78, 89, 95, 117–18, 124, 151. *See also* Berlin; Hamburg; Hanover; Munich
glanders, 111–13, 128
Golitsyn, Vladimir, 43
Gorbunov, Dmitrii, 124, 141
Gornostaev, Ivan, 159
Gortynskaia, Olga, 186
grazing, 80, 104–5, 110, 113
Great Reforms, 14, 19, 27, 30, 117, 133
guardianships of the poor, 164–67
Gurin, Gavriil, 135, 138–39

Hamburg, 44, 54, 90, 125, 155, 157
Hanover, 122
Helmann, Christopher, 112
Hobrecht, James, 62–64, 71
horses, 104–5, 112, 222n5, 226n10
hospitals: Alekseev psychiatric hospital, 35, *37*, 49; food at hospitals, 82–83; for newborns and children, 4, 36, *48*, 163–64, 184, 191, 207; for venereal disease, 161; state hospitals, 26, *47*, 161; in the early twentieth century, 47–49, 164, 203–7; under Alekseev, 35–37, 41, 161; zemstvo hospitals, 28
housing: and access to sewerage, 74–78; at the abattoir, 141–44; at the sewage farm, 83; construction material, 71–72; of the poor, 21, 151, 159–61, 167–72; prices, 72; sanitary control, 39
hygiene: as science, 19–26; and children, 151–52, 157, 161, 171; and community medicine, 27–30. *See also* school hygiene

illegitimate children, 153–55
imperial administration, 26, 32–36, 40–46, 90–91, 169, 203
industrial waste, 25, 74, 83, 87–98
infant care, 4, 149–57, 161–67. *See also* mortality
Institute of Hygiene, 21, 23, 213n16
intelligentsia, 18, 25, 35

Kalning, Otto, 112
Kastalskii, Vsevolod, 64–66
Khitrov market, 72, 167, 169
Kishkin, Nikolai, 167
Koch, Robert, 23, 90, 112

laboratory, 8, 21–23, 40, 60, 117, 125, 129–30
Lanin, Nikolai, 121
Lebedev, Ivan, 187
legislation: on factories, 86, 150; on children, 150; on meat, 126–27; on water pollution, 86–89, 94–97; sanitary, 25. *See also* industrial waste
liberalism, 25, 33–34, 45
livestock census, 222n5
livestock trade, 102–10, 113
Loeffler, Friedrich, 112
London: gender composition, 17; infant mortality, 155, 157; population density, 159; sewage system, 54, 64; slaughterhouses, 122, 124; typhoid deaths, 61

mallein test, 112–13
maternity leave, 161, 167, 171
measles, 60, 160
meat: as a source of health, 101–2, 197; at schools, 187–88; at summer colonies, 197; ethical concerns, 134, 136; quality control, 125–30; supply to Moscow, 103–8, 207

Index

Mechnikov, Ilya, 23
Medical Council, 93–94, 118, 167
medical education, 19–20
medical examination at schools, 184, 186–87, 189, 191–94, 197
Medical Statute, 86
medicalization, 114, 151, 161, 180, 193, 198, 202–3
Medical-police committee, 40
Memphis, Tennessee, 9, 55, 64–65
merchants, 17, 32–35, 43, 168, 183
miasmatic theory of disease, 9, 22, 53–54, 59–60, 65, 68
Mikhailov, Nikolai, 184, 186–87, 192, 197
milk, 103, 152, 160, 187–88, 197, 200. *See also* breastfeeding; milk kitchens
milk kitchens, 164, 171, 207
Ministry of Finance, 26, 44, 88, 90, 95
Ministry of Internal Affairs, 20, 26
Ministry of Public Instruction, 20, 26, 43, 181–82
Ministry of Trade and Industry, 94–96
Molleson, Ivan, 152
mortality: infant, 4, 149–57, *156*, 160, 163, 170–71; general, 7, 50, *156*, 157, 170; from infectious diseases, 60, 160, 171
Moscow, *15*; demographics, 14–17, 58, 65, 70, 206; electoral system, 18, 42–43, 48; plan of, *16*; social composition, 16–19; social topography, 71–72
Moscow Exchange Committee, 94–95
Moscow Imperial Foundling Home, 153–57, 166, 170
Moscow Society for the Diffusion of Scientific Knowledge, 64, 66
Moscow University, 20–21, 23, 43, 60, 128, 149

Moskva River, 49, *55*, 56–57, 63, 71, 78–79, 91–93, 119
Munich, 21, 155
Municipal Statute, 18–19, 42, 169, 201
municipalization, 6, 35–36, 42, 56, 64, 67–68, 202–3

Nagorskii, Valentin, 41, 122, 129, 176
Naumov, Aleksandr, 101
Naumov, Dmitrii, 28
Neva, 57
Nicholas I (1825–1855), 33
Nicholas II (1894–1917), 43, 46, 95, 208, 227n20
night shelters, 75, 167–68, 184
Nikolai Nikolaevich the Elder, Grand Duke, 133
nomads, 104–6
nurseries, 152, 161, 164–68, *165*, 171, 205
nutrition, 24–25, 151, 157, 160, 187–88, 196–97

Okunev, Nikita, 208–9
Osipov, Evgraf, 28
Ottoman Empire, 21, 104
outpatient clinics: general, 37, *38*, 40–41, 48–49, 157, 163–64, 171, 203; for animals, 113; for school children, 174, 191, 199–200

Paris, 17, 54–55, 57, 61, 122, 157, 159
Pasteur, Louis, 23, 67
pastures, 79–80, 82, 103, 105–7, 117
Pavlov, Illarion, 177
peasants: in Moscow, 16–17, 153, 155, 157, 183; in the city government, 43; and the Moscow sewage system, 78–84; and livestock trade, 103, 106–7, 109, 112
Petrograd. *See* Saint Petersburg

Index

Pettenkofer, Max von, 21, 23
physical education, 25, 178, 191
Pirogov Society of Russian Physicians, 21–22, 111
playing, 178, 191, 193, 196, 199–200
police, 26, 39–40, 43–45, 85–86, 91–96, 109–11
Popov, Mikhail, 61–64, 66–68, 71
prasols, 106–7
priests, 181, 183, 187
prostitution, 39–40, 113, 150
provincial medical boards, 26, 94
psychiatry, 49, 150, 185, 193, 198. *See also* hospitals
public good, 35, 42, 49, 81–84, 133, 143–45

rabies, 111, 112, 128
railways, 46, 109–10, 117, 119, 124, 131, 207
Rein, Georgii, 167
respiratory diseases, 160, 163
Revolution of 1905, 19, 43–44, 46–48, 143–46, 167, 204
Revolution of 1917, 3–4, 71, 84, 96, 171, 208
Richter, Nikolai, 183–84
rinderpest, 108–13, 117–18, 129
Romanov, Sergei. *See* Sergei Alexandrovich, Grand Duke
Rubinstein, Nikolai, 32
Rudnev, Mikhail, 117
Rukavishnikov, Konstantin, 43
Russian Society for the Protection of Animals, 9, 133–39, 226n10
Russian Technical Society, 63, 159

Saint Petersburg: abattoir, 116, 119, 122, 145; cholera deaths, 90; foundling home, 154; infant mortality, 159; meat supply to, 105–9; school system, 183; sewerage, 54; trichinosis, 117; typhoid deaths, 61. *See also* Neva
sanitary bureau, 40
sanitary congresses, 111
sanitary engineering and technology, 69, 120, 122, 157, 169, 202–6
sanitary inspection, 31, 36, 41, 170, 192, 203, 207. *See also* school sanitary inspection
sanitary-bacteriological synthesis, 24, 26
scabies, 112, 192
scarlet fever, 60
school doctors. *See* school sanitary inspection; summer colonies for weak children
school hygiene, 174–80, 196. *See also* school sanitary inspection
school sanitary inspection, 174, 186–94, 203, 205, 207
schools: buildings, 177, 180, 188–89, *190*, 191, 194; curriculum, 178, 191; enrollment, 173, 181–83
Schütz, Wilhelm, 112
Sergei Alexandrovich, Grand Duke, 43–46, 85, 91, 127
sewage farming, 62, 66, 78
sewage system: access to, 71–78; combined vs. separate system, 64–69; overflows, 64–67; plan of Moscow's, *77*
sewage treatment. *See* waste treatment
Shchepkin, Mitrofan, 201–2
sheep, 104–5, 222n5
Shervud, Vladimir, 67–68
Shone, Isaac, 125
Siegerist, Henry E., 43
slaughterhouse reform, 115, 119–24
smallpox, 28, 36, 60, 184, 191, 194, 199

Index

smell, 14, 53–54, 58–61, 93, 118–19, 122, 131
social insurance, 167
Sokolov, Andrei, 60
Stanislavsky, Konstantin, 32
State Duma, 19, 46, 48, 96, 182
statistics: housing, 159; mortality, 7, 10, 151–57, 161, 163, 205; responsibility for, 26, 39; veterinary, 114, 126; schooling, 173
Statute of Industry, 86
Statute of Magistrates, 87, 133
strikes, 91, 138, 143–44
Sumbul, Leonid, 121
summer colonies for weak children, 194–200, *195*
Suslova, Nadezhda, 20
swine erysipelas, 112
syphilis. *See* venereal disease
Sysin, Aleksei, 204, 209–10

Tchaikovsky, Piotr, 32
toilets, 37, *75*, 76, 142, 146, 160, 177, 189
Tolstoy, Leo, 134
transnational municipalism, 7, 14, 29
Trepov, Dmitrii, 44–46, 85, 91–92, 216n40
trichinosis, 117–18, 125
tuberculosis, 112, 160, 176
typhoid, 60–61, 67
typhus, 36, 60, 208–9

Union of Towns, 206–7
Union of Zemstvos, 206
United States, 3–4, 65–69, 78, 89, 110, 133, 191

vaccination, 28, 112, 184, 191, 194, 199–200, 232n12
vegetarianism, 134

venereal disease, 33, 39–40, 161, 170
Verderevskii, D., 135–39
veterinary medicine, 26, 41, 108–14, 117–18, 122, 128–30, 202
veterinary organization, 39, 111–13, 125, 128–30, 135, 139–40
Vienna, 17, 54, 122, 125, 129
Virchow, Rudolf, 117, 175
Virenius, Aleksandr, 180
Volga, 56, 104

Waring, George, 9, 64–65, 68
waste treatment: at the abattoir, 119–20; industrial, 88–89, 93–98, 204; sewage, 53, 59, 65–69, 74, 78–84, 203
water closets. *See* toilets
water quality standards, 89, 92–94, 97–98
Waterways Protection Committee, 95–96, 98
weight gain: animals, 105, 110, 113–14; children, 197, 200
Winslow, Charles-Edward Emory, 3–4, 171, 207, 211n6
Witte, Sergei, 44, 88, 92
women: as physicians, 40, 174, 186–87; at schools, 183; at the abattoir, 140, 142; in Moscow, 17–18, 152–53, 155, 157, 164, 166, 168. *See also* prostitution
workers: and elections, 18, 43, 46; at the abattoir, 135, 138–46; at factories, 58, 72, 86, 91, 93, 98, 150; at the sewage farm, 83; in the province of Moscow, 28; migrant, 153, 159, 167, 183

Yauza, 56–58, *57*, 60
yellow tickets, 40

Zelenin, Nikolai, 129
zemstvo: and childcare, 150, 166–67; and schools, 173–74, 176, 179, 184, 195; medicine, 3, 7–9, 25–30, 203–4 (*see also* community medicine); Moscow, 21, 28–29, 32, 41, 79–81, 94; political importance, 19, 27, 36, 96, 206; veterinary medicine, 41, 109, 111–12
zemstvo sanitary model, 8–9, 31, 41, 204
Zhbankov, Dmitrii, 27
zoonosis, 39, 41, 111, 205
Zubatov, Sergei, 91, 216n40
Zurich, 20, 43